Computational Text Analysis for Functional Genomics and Bioinformatics

Computational Text Analysis for Functional Genomics and Bioinformatics

Soumya Raychaudhuri

OXFORD
UNIVERSITY PRESS

OXFORD
UNIVERSITY PRESS

Great Clarendon Street, Oxford OX2 6DP

Oxford University Press is a department of the University of Oxford.
It furthers the University's objective of excellence in research, scholarship,
and education by publishing worldwide in

Oxford New York

Auckland Cape Town Dar es Salaam Hong Kong Karachi
Kuala Lumpur Madrid Melbourne Mexico City Nairobi
New Delhi Shanghai Taipei Toronto

With offices in

Argentina Austria Brazil Chile Czech Republic France Greece
Guatemala Hungary Italy Japan Poland Portugal Singapore
South Korea Switzerland Thailand Turkey Ukraine Vietnam

Published in the United States
by Oxford University Press Inc., New York

First published 2006

British Library Cataloguing in Publication Data
Data available

Library of Congress Cataloging in Publication Data
Data available

Typeset by SPI Publisher Services, Pondicherry, India
Printed in Great Britain
on acid-free paper by
Biddles Ltd., King's Lynn Norfolk

ISBN 0-19-856740-5 978- 0-19-8567400
ISBN 0-19-856741-3 (Pbk.) 978-0-19-8567417 (Pbk.)

1 3 5 7 9 10 8 6 4 2

Dedicated to my grandfather and role model
Professor Sahadeb Banerjee (4/1/914–4/20/2005)

Preface

This book is an introduction to the newly emerging field of textual analysis in genomics. It presents some of the newest methods, and demonstrates applications to proteomics, sequence analysis, and gene expression data.

My personal interest in this field began early during my graduate school years as these methods were rapidly emerging. My colleagues were excitedly utilizing new high throughput technologies in biology with which they could collect data at unprecedented rates. Gene expression arrays, for example, offered the opportunity to simultaneously explore expression of all genes in a cell. However, many were hitting the same roadblocks; making sense of all of that data was tedious and frustrating. Even differentiating signal from noise was a challenge; certainly finding subtle patterns in the data proved to be much more difficult than anyone expected. A host of statistical methods were emerging to analyze the numerical data, but yet they lacked the necessary context to fully harness the power of these complex experimental results. The difficulty is that complete interpretation requires understanding all of the large number of genes, their complex functions, and interactions. But, just keeping up with the literature on a single gene can be a challenge itself, and for thousands of genes it is simply impossible! At that time I became interested in the promise of statistical natural language processing algorithms, and their potential in biology. These methods often are the only reasonable way to include the literature on thousands of genes in genomics data analysis and to give context to the data.

We describe analytical methods that utilize the scientific literature in the context of specific experimental modalities in this book. But much of what is discussed here can easily be generalized to most large-scale experimental methods. For example, the expression array methods can be generalized to any numerical data set, and the protein interaction methods can be generalized to any type of interaction. In addition to devising the theory behind the methods, we emphasize real world examples and evaluations in this book to demonstrate how methods can be applied practically and what performance benefit they offer.

This book can be used as a primary text in a graduate course in a genomics or computational biology curriculum, or as an adjunct text in an advanced computational biology course. The book has been written with sufficient background material and the prerequisites for this book are few. A basic understanding of probability and statistics is helpful at the level of an introductory undergraduate course. Basic biological and bioinformatics concepts are reviewed to the extent that is necessary. No background in computational text analysis is necessary, but is certainly helpful.

We are hopeful that this text will encourage the reader to develop and utilize these methods in their own work, and to maximize the potential of large-scale biology.

Acknowledgements

This book was to a large extent the product of work that I started under the guidance of Russ Altman, who has mentored me through the years. In addition, Jeffrey Chang and Hinrich Schutze have been great influences in these pursuits. Patrick Sutphin, Farhad Imam, Joshua Stuart, Nipun Mehra, Amato Giaccia, Peter Small, and David Botstein are all colleagues that have influenced and shaped the content of this book. It has been a pleasure working with Alison Jones and her associates at Oxford University Press. Sourobh Raychaudhuri, my brother, has given me feedback on specific sections. Finally, I thank Meenakshy Chakravorty, my wife, whose critical suggestions on this manuscript have been invaluable.

Soumya Raychaudhuri
Boston, USA, 2005

Contents

List of Figures

List of Plates

(At End)

List of Tables

1 An introduction to text analysis in genomics

The February 16th, 2001 issue of *Science* magazine announced the completion of the human genome project—making the entire nucleotide sequence of the genome available (Venter, Adams et al. 2001). For the first time a comprehensive data set was available with nucleotide sequences for every gene. This marked the beginning of a new era, the "genomics" era, where molecular biological science began a shift from the investigation of single genes towards the investigation of all genes in an organism simultaneously.

Alongside the completion of the genome project came the introduction of new high throughput experimental approaches such as gene expression microarrays, rapid single nucleotide polymorphism detection, and proteomics methods such as yeast two hybrid screens (Brown and Botstein 1999; Kwok and Chen 2003; Sharff and Jhoti 2003; Zhu, Bilgin et al. 2003). These methods permitted the investigation of hundreds if not thousands of genes simultaneously. With these high throughput methods, the limiting step in the study of biology began shifting from data collection to data interpretation. To interpret traditional experimental results that addressed the function of only a single or handful of genes, investigators needed to understand only those few genes addressed in the study in detail and perhaps a handful of other related genes. These investigators needed to be familiar with a comparatively small collection of peer-reviewed publications and prior results. Today, new genomics experimental assays, such as gene expression microarrays, are generating data for thousands of genes simultaneously. The increasing complexity and sophistication of these methods makes them extremely unwieldy for manual analysis since the number and diversity of genes involved exceed the expertise of any single investigator.

The only practical solution to analyzing these types of data sets is using computational methods that are unhindered by the volume of modern data. Bioinformatics is a new field that emphasizes computational methods to analyze such data sets (Lesk 2002). Bioinformatics combines the algorithms and approaches employed in computer science and statistics to analyze, understand, and hypothesize about the large repositories of collected biological data and knowledge.

However, the most critical resource of relevant information necessary for genomic data interpretation is the peer-reviewed published literature about the individual genes. While its value is without question, incorporating it into large-scale computational analyses is challenging. Text is available in large quantities, but is often disorganized and contradictory. In addition, accurate computation on text is a challenging subject.

This book introduces automatic algorithms to access and utilize the intractably vast sources of literature to help interpret such large data sets. As documents become available in electronic format, and they become necessary for genomic-scale biology, bioinformatics researchers are investigating the application of natural language processing methods to mine biological text (Yandell and Majoros 2002; Shatkay and Feldman 2003). This is an area of growing interest—with dozens of active groups and internationally publicized symposia. While the content of biomedical literature is undeniably valuable, it is poorly structured for computation (compared to databases, for example). Fortunately, the field of natural language processing has been pursuing techniques to understand and interpret unstructured text (Manning and Schutze 1999). The field of text mining and computational natural language processing is a well established one that has made many advances and gains over the past decades. Many of the techniques from that field can be applied directly to problems in genomics. However there are many more challenges and opportunities as well, since the challenge in bioinformatics is not just to mine the literature, but also to utilize it to understand experimental data.

1.1 The genomics literature

The biological literature is a vast and comprehensive resource. Every accepted facet of biological knowledge is locked in the published literature. Investigators in genetics and genomics strive to publish their results in reputable journals. In Table 1.1 we have listed some of the best-known journals. Publication typically requires a thorough review by scientific peers to assess the originality and validity of the purported results and interpretation. As a result, peer-reviewed publications are a relatively reliable resource of information. Most papers today report experimental results on a single gene or protein, though more and more large-scale experiments are being reported. Increasingly, the scientific literature is becoming available online in electronic format, raising the possibility of facile computational analysis. PubMed abstracts are increasingly available for most biologically relevant articles (see Plate 1.1 and Figure 1.1), while electronic publishers

Table 1.1 *Popular molecular biology journals.* The above is a list of some of the best-known journals that contain articles relevant to genetics and molecular biology. Many of the articles published in these journals are pertinent to the study of specific genes. We also list the number of articles published by these journals in 2003 in the right column.

	Journal	Total Articles 2003
1	American Journal of Human Genetics	330
2	Cancer Cell	134
3	Cancer Research	1311
4	Cell	356
5	Developmental Cell	208
6	European Molecular Biology Organization Journal	653
7	Genes and Development	288
8	Genome Research	291
9	Human Molecular Genetics	390
10	Immunity	173
11	Journal of Cell Biology	457
12	Journal of Experimental medicine	493
13	Molecular Cell	325
14	Molecular Cell Biology	803
15	Nature	2408
16	Nature Cell Biology	238
17	Nature Genetics	308
18	Nature Immunology	236
19	Nature Medicine	412
20	Nature Neuroscience	264
21	Nature Structure Biology	210
22	Neuron	421
23	Plant Cell	256
24	Proceedings of the National Academy of Science USA	2955
25	Science	2377

such as High-Wire press and PubMed Central permit access to full text articles (Hutchinson 1998; Roberts 2001). Several of the newer journals publish strictly online; the Public Library of Science (PLOS) and BioMed Central are two publishers that produce open access peer-reviewed journals online. The wide availability of papers in electronic format makes computational analysis for genomic data analysis feasible.

The quantity of available literature is vast. For example, as of January 2005 the PubMed database contains some 14 million biomedical abstracts from over 30,000 journals published in the last 50 years. As of June 2004, SwissProt, a protein sequence database, has references to 95,654 articles from 1523 journals that are assigned to a total of 137,095 protein sequences (Boeckmann, Bairoch et al. 2003). As of June 2004, LocusLink, a curated genomic data resource, contains references to some

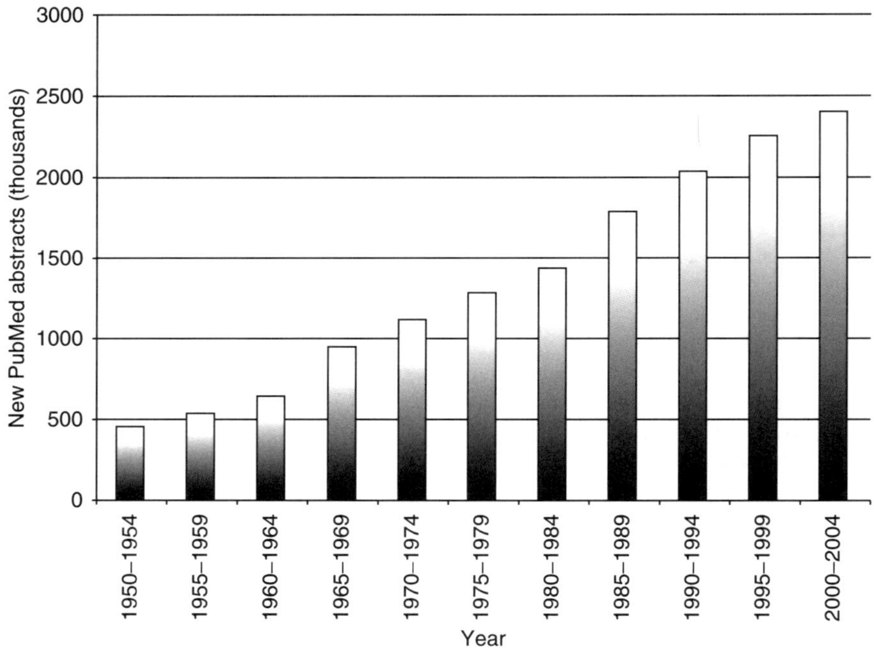

Figure 1.1 *PubMed abstracts.* New PubMed abstracts added to the PubMed database are depicted below as a function of time. The number of biological abstracts submitted has been rapidly increasing over the last 50 years. This plot was derived from publication dates listed in PubMed records.

121,557 articles that are pertinent to 96,577 specific genes (Pruitt and Maglott 2001).

Besides the inherent challenges of text mining, the genomics literature has other issues that make computation difficult. To begin with, all documents do not have equal value by any means. Scientists generally regard some papers to be more credible and to have more value than others. As a result, most fields have a few extremely well regarded papers, whose results have been replicated and are well cited. But the majority of papers do not have similar wide regard, and they are cited more sparsely. Assessing the quality or reliability of a given paper is a challenge. Manual evaluation by independent reviewers, such as the Faculty of 1000 (www.facultyof1000.com), is a thorough but tedious option. The quality of the journal, the number of citations, the reputation of the author, and the credibility of the publishing institution, are all factors that can be used as heuristics to assess the importance of a given paper.

In addition, even if the quality of a paper can be judged computationally in an absolute sense, there is a greater difficulty in assessing the relevance of a paper for the specific task at hand. For example if we are creating an algorithm that automatically reads and assigns the cellular function of a gene from a document, we must insure that the document, besides being a quality publication, is pertinent to the gene. In addition, we must be careful to ascertain whether the document's focus is pertinent to the gene's cellular function and not other aspects of the gene, such as medical diseases that the gene has been implicated in.

The diversity of subjects addressed in the biological literature is enormous. A single gene may have references that discuss the function of the gene at a molecular level, the larger cellular system the gene is involved in, the role the gene plays at the whole-organism level, diseases that a gene is involved in, the structure of the protein product of a gene, or different mutant forms of the same gene. In addition, while most articles are very scientifically specific, others are very broad. For example, a review article about a particular gene may focus on a broad array of subjects while another scientific article may address a very specific aspect of the same gene. Similarly, while many articles address individual genes, others may be germane to large numbers of genes. Of the 121,577 LocusLink references, 80,724 refer to only a single gene, while 217 refer to over 50 genes. The skewed distribution of the number of genes referred to by different articles is displayed in Figure 1.2.

An additional difficulty is that the availability and quality of literature is very biased towards well-studied areas. For example, some genes are extremely well studied, but many have not been studied at all. If we consider human genes, as of June 2004 there are 36,857 genes listed in the LocusLink database, and of these genes only 22,489 have references associated with them; so a full one-third of known genes do not have a single relevant published paper. At the other extreme are genes such as the *tumor protector p53* gene, which has some 580 relevant articles listed in LocusLink, and *vascular endothelial growth factor*, which has 277 listed. The skew in the distribution of genes is apparent in Figure 1.3. Any text mining algorithms that are used in the field of genomics must take into consideration the inequities of the published literature.

1.2 Using Text in genomics

There are certainly many areas in genomics where the use of scientific text can be invaluable. In this section we give a short description of three areas where text mining might augment genomics analysis.

1.2.1 Building databases of genetic knowledge

One of the active areas of interest in the field is the use of text mining strategies to scan the literature and summarize pertinent facts in a

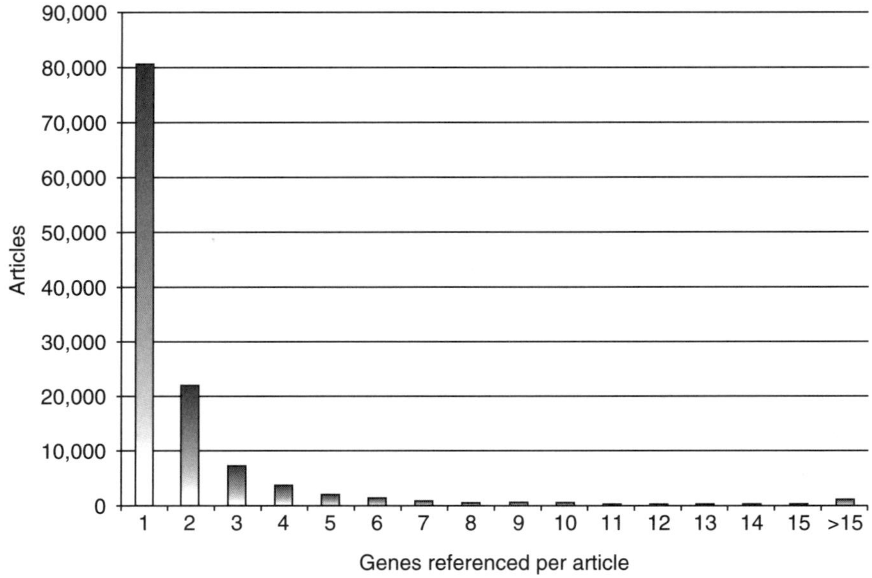

Figure 1.2 *Distribution of articles as a function of the number of genes referenced.* Here we have plotted the distribution of articles indexed in LocusLink as a function of the number of genes that they reference. It is a highly skewed distribution. While the vast majority of articles are specific in subject matter and refer to only a few genes, a few refer to a large number of genes.

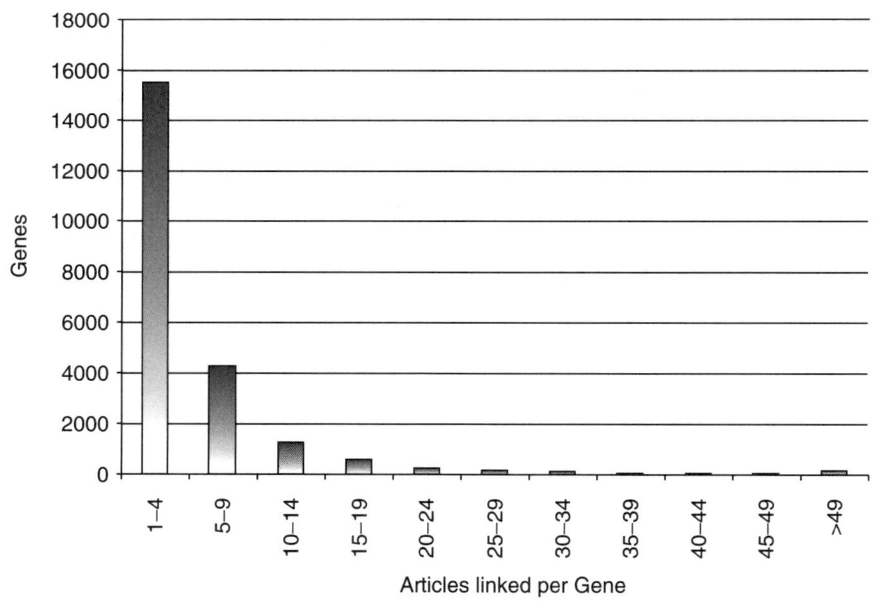

Figure 1.3 *Distribution of genes as a function of available relevant articles.* Here we have plotted the distribution of human genes listed in LocusLink as a function of the number of articles that pertain to them. It is a highly skewed distribution. While the vast majority of genes are relatively poorly studied genes with less than five articles, a few genes are extremely well studied with many articles.

structured format in databases or knowledge bases (Yandell and Majoros 2002; Shatkay and Feldman 2003). Genetic knowledge from the literature is more appropriate for computation once it has been structured properly. With the advent of genomics, online resources that contain structured information about genes, such as gene function, gene interactions, and protein product interactions, have become extremely valuable tools to interpret large data sets. For example, The Gene Ontology (GO) consortium and the Munich Information Center for Protein Sequences (MIPS) provide vocabularies of function and assign genes from multiple organisms the relevant terms (Ashburner, Ball et al. 2000; Mewes, Frishman et al. 2000). Similarly, the Database of Interacting Proteins catalogs protein interactions (Marcotte, Xenarios et al. 2001).

However, such resources may not be comprehensive and up to date at any given time, and it is also laborious to maintain the vocabulary and the gene assignments. Currently, most resources are maintained and expanded by careful manual interpretation. Text mining offers an opportunity to create algorithms that could automatically and rapidly scan the large volume of literature, as well as new papers as they become available, and then enter their information content into databases.

This approach has been proposed for databases of biological function and for databases that keep track of interactions between genes (Ohta, Yamamoto et al. 1997; Fukuda, Tamura et al. 1998; Sekimizu, Park et al. 1998; Tamames, Ouzounis et al. 1998; Blaschke, Andrade et al. 1999; Craven and Kumlien 1999; Ng and Wong 1999; Humphreys, Demetriou et al. 2000; Proux, Rechenmann et al. 2000; Thomas, Milward et al. 2000; Marcotte, Xenarios et al. 2001; Ono, Hishigaki et al. 2001; Stephens, Palakal et al. 2001; Wong 2001; Raychaudhuri, Chang et al. 2002; Donaldson, Martin et al. 2003). This area is a challenging one, however, as effective algorithms must be available that can (1) identify entity names, such as gene or protein names in the literature with high accuracy, (2) recognize if a relevant function or interaction is being described, and (3) classify the description into one of potentially a great many functions and interactions. All of these tasks are challenging in themselves and require high levels of accuracy.

1.2.2 Analyzing experimental genomic data sets

Designing bioinformatics algorithms that analyze genomic data sets is a significant problem. The challenge in writing such methods is not only in effectively interpreting the complexity for such data sets, but also overcoming the noise that is inherent in many of these experimental approaches. One common solution to addressing the challenges of interpreting genomic data sets is using external data from other sources to guide or verify possible analytical results. For example, if the interpretation of an experiment is consistent with another independent experiment our confidence in that interpretation is increased. One vast and very good source of external information is the corpus of peer-reviewed publications.

The general approach is, given a genomics data set, to generate many possible hypotheses that might explain the observed result. Then compare the hypotheses to what is reported in the large body of the scientific literature, and then to pick the hypothesis that is most consistent with what has been reported (see Plate 1.2). In a sense this is very similar to the normal paradigm under which scientific exploration is undertaken. Typical interpretation of a scientific result requires the investigator to examine prior results to pick an interpretation that is most consistent. This general approach has been applied to gene expression data analysis and sequence analysis with success (MacCallum, Kelley et al. 2000; Shatkay, Edwards et al. 2000; Chang, Raychaudhuri et al. 2001; Jenssen, Laegreid et al. 2001; Raychaudhuri, Chang et al. 2003). In the case of sequence analysis, alignment methods may suggest possible families of related proteins; the validity of these families can be checked against the scientific literature. In the case of gene expression analysis, the data may suggest co-regulation of a cluster of genes; the possibility of co-regulation can then be checked against the literature.

1.2.3 Proposing new biological knowledge: identifying candidate genes

One of the great hopes of the genomic revolution is that the fruits of our labor might transfer into medicine in the form of a better understanding of human disease. In the investigation of particular diseases, many investigators choose to select a large set of genes, and then conduct specific genetic studies to identify the ones that are truly involved in a particular disease. One of the challenges to these approaches is the selection of the initial set of genes for investigation. It is undesirable to cast too wide a net and select too many genes, as the number of false positives will be high. On the other hand, selecting too few genes risks losing the genes that might be most critical to the particular disease.

Mining the biological literature can offer some insight to this problem. Currently genes are picked by expert knowledge. Instead, text-mining strategies can be used to identify genes that have potential associations with the particular disease or concepts that are associated with these genes. At the time of writing, this is a completely uninvestigated area of research. We are hopeful that in the coming years this area will be thoroughly addressed.

1.3 Publicly available text resources

1.3.1 Electronic text

As mentioned above, there are an increasing number of primary resources from which the text of articles is readily available online for analysis. The most important for text mining in genomics so far has been PubMed (www.ncbi.nlm.nih.gov/PubMed/), a National Institute of Health funded database of article abstracts (see Plate 1.1). Almost every biologically significant document published in the last 50 years has its abstract available in PubMed. The articles are in Medline format, a standardized format with fields for the authors, abstract, and keywords, and other pertinent details (see Figure 1.4). Included in the format are the MeSH headings; these are keywords assigned from a hierarchical controlled vocabulary by experts who have read the original article; these headings can be very useful in searching for articles.

PubMed Central (www.pubmedcentral.nih.gov) is a newer database of whole text articles (Roberts 2001). As of June 2004 the whole text of about 285,000 individual articles from 152 different journals is available without charge. Most of the articles have been published after 1990. This is a very exciting new initiative for genomics and text mining as it permits the possibility that algorithms can be produced to mine other sections of an article rather than just the abstract. Whole-text mining also offers the possibility of accessing the valuable information that is contained in the article figures (Liu, Jenssen et al. 2004). Hopefully, in the future initiatives such as PubMed Central will become more comprehensive.

In addition whole text is becoming available from individual publishers as well. Publishers such as Biomed Central (www.biomedcentral.com) and the Public Library of Science (www.publiclibraryofscience.org) are creating reputable journals that are published exclusively online. Other traditional journals are now publishing articles online as well. Highwire (www.highwire.org) press has made over 700,000 articles available from 362 journals.

1.3.2 Genome resources

Many of the publicly funded genome databases have reference indices that are valuable for text mining. Databases such as FlyBase (flybase.bio.indiana.edu), Wormbase (www.wormbase.org), Mouse Genome Database (MGD) (www.informatics.jax.org/mgihome/MGD/aboutMGD.shtml), and Saccharomyces Genome Database (SGD) (www.yeastgenome.org) all have indices that link article references to genes (Cherry, Adler et al. 1998; Stein, Sternberg et al. 2001; Blake, Richardson et al. 2002; FlyBase 2002). Most of the links are to PubMed abstracts. These indices are usually manually derived and updated,

```
PMID- 8976574
OWN - NLM
STAT- MEDLINE
DA  - 19970319
DCOM- 19970319
LR  - 20041117
PUBM- Print
IS  - 0961-8368
VI  - 5
IP  - 12
DP  - 1996 Dec
TI  - Insights into the local residual entropy of proteins provided by NMR
      relaxation.
PG  - 2647-50
AB  - A simple model is used to illustrate the relationship between the dynamics
      measured by NMR relaxation methods and the local residual entropy of
      proteins. The expected local dynamic behavior of well-packed extended
      amino acid side chains are described by employing a one-dimensional
      vibrator that encapsulates both the spatial and temporal character of the
      motion. This model is then related to entropy and to the generalized order
      parameter of the popular "model-free" treatment often used in the analysis
      of NMR relaxation data. Simulations indicate that order parameters
      observed for the methyl symmetry axes in, for example, human ubiquitin
      correspond to significant local entropies. These observations have obvious
      significance for the issue of the physical basis of protein structure,
      dynamics, and stability.
AD  - Department of Biological Sciences, State University of New York at Buffalo
      14260, USA. wand@jasper.chem.buffalo.edu
FAU - Li, Z
AU  - Li Z
FAU - Raychaudhuri, S
AU  - Raychaudhuri S
FAU - Wand, A J
AU  - Wand AJ
LA  - eng
GR  - DK-39806/DK/NIDDK
PT  - Journal Article
PL  - UNITED STATES
TA  - Protein Sci
JID - 9211750
RN  - 0 (Proteins)
SB  - IM
MH  - Entropy
MH  - Humans
MH  - Magnetic Resonance Spectroscopy
MH  - Models, Theoretical
MH  - Proteins/*chemistry
MH  - Research Support, U.S. Gov't, P.H.S.
EDAT- 1996/12/01
MHDA- 1996/12/01 00:01
PST - ppublish
SO  - Protein Sci 1996 Dec;5(12):2647-50.
```

Figure 1.4 *Article in MedLine format.* MedLine format is the document format that PubMed abstracts are stored in. Fields include the authors (AU), their affiliations (AD), reference number (PMID), abstract (AB), title (TI), language (LA), information about research funding (GR), the type of publication (PT), journal title (TA), and MeSH keyword information (MH).

so in general they are very high quality and reliable. Summary statistics for the reference indices are presented in Table 1.2. The number of genes that have article references and the number of article references listed in the reference index are very dependent on how well the organism in question has been explored. In addition genome databases often have other valuable information such as keywords indicating the function of different genes, gene nucleotide sequences, and protein product amino acid sequences. These reference lists are extremely useful for most applications of text mining to genomics.

Table 1.2 *Reference indices from different genomic resources.* Recent statistics for six genomic database reference indices. All have thousands of articles that refer to thousands of genes. In all cases the mean number of gene references per article exceeds the median; this is because the distribution is skewed with a few extreme articles referring to many genes. Similarly, the mean number of articles linked per gene exceeds the median since there a few very well studied outlying genes with many germane articles. Data about LocusLink and SwissProt are from June 2004. Data about the other resources are from March 2002.

	SGD	MGD	Flybase	Wormbase	LocusLink	SwissProt
Organism	Yeast	Mouse	Fly	Worm	Many	Protein Sequences
Genes with article references	5151	26,148	14,732	2289	121,577	137,095
articles	22,934	41,669	15,495	2144	96,577	95,654
Number of genes referenced per article median	2	1	3	4	1	1
mean	2.73	2.73	6.27	6.37	3.44	2.63
Number of article linked per gene median	4	1	1	2	2	1
mean	12.12	4.35	6.59	5.97	4.33	1.83

LocusLink (www.ncbi.nih.gov/LocusLink/) and its successor Entrez Gene are resources that integrate genetic information from 15 different species, ranging from viruses to mammals (Pruitt and Maglott 2001). It too has a comprehensive reference index that associates genes with PubMed abstracts. Its summary statistics are also presented in Table 1.2. It includes a very large number of genes and article abstracts.

Curators of the biological data repositories, such as Swiss-Prot, often provide links from individual data records to PubMed abstracts (Boeckmann, Bairoch et al. 2003). For example, a protein sequence submitted to the Swiss-Prot database might be linked to the PubMed abstract of the article with details about the protein. Summary statistics for the Swiss-Prot reference index are provided in Table 1.2.

Countless other databases are available online that have reference indices that suit their specialized need.

1.3.3 Gene ontology

Another critical resource in functional genomics is the Gene Ontology (GO). We mention it here because of the special significance that it has in this book (Ashburner, Ball et al. 2000). The Gene Ontology is a controlled vocabulary of terms that describe gene function. The terms are organized into three broad branches. "Molecular Function" terms describe the biochemical reactions that a protein catalyzes. The "Biological Process" terms describe the global physiologic process that a protein is involved in. "Cellular Location" terms describe the compartment of a cell that a protein product is situated in. A properly annotated gene may have multiple terms from each of these branches of Gene Ontology.

Gene Ontology literally contains thousands of terms that describe genetic attributes ranging from very broad terms ("*metabolism*") to very specific ("pyruvate kinase"). One of the very special and valuable features of Gene Ontology is that it is organized hierarchically. More specific functional terms are children of more general terms. So if a gene is assigned a particular specific Gene Ontology term, it will have the function associated with that term, and also will have the functions described by the parent terms as well. For example the term *glycolysis* would be a descendent of the term *carbohydrate metabolism*. All genes assigned the term *glycolysis* are by default involved in *carbohydrate metabolism*. However, all *carbohydrate metabolism* genes are not necessarily *glycolysis* genes. So, depending on the current state of knowledge, annotations for a gene can be made as specific or general as necessary.

Currently the GO consortium is annotating genes from many different organisms with GO terms. The present state of Gene Ontology annotations is described in Table 1.3. In Chapter 8 there is a more detailed discussion of Gene Ontology and gene annotation.

1.4 The advantage of text-based methods

At present, text-based analytical approaches are mostly an emerging technology. Most of the methods presented in this book are in the preliminary stages. Pilot studies have demonstrated encouraging results; wide-scale application is still pending.

However these methods are promising. As discussed above there are excellent and easy to access literature resources available that are ever expanding. As we will demonstrate in the coming chapters, inclusion of literature in genomics data analysis algorithms has in some cases already demonstrated the discovery of novel biology! The role of text in genomics data analysis is particularly effective in cases where the data quality is poor. For example in expression data or yeast-2-hybrid data, the scientific text can very effectively help in the separation of valuable biology from noise. Inclusion of automated text analysis can offer avenues to make sense of noisy data.

Under all circumstances, however, inclusion of text can result in a great speedup in the analysis of data, and hence in scientific discovery. For

Table 1.3 *GO annotations*. This table lists some of the organisms whose genes have been assigned Gene Ontology terms. In the second column the source database is listed. In the third column the total number of annotated genes is listed. In the final column the total number of article references used to generate some of those annotations is listed.

Organism	Database	Genes with GO annotations	Article references
Saccharomyces cerevisiae	SGD	6459	4849
Drosophila melanogaster	FlyBase	9538	6715
Mus musculus	MGI	15380	4725
Arabidopsis thaliana	TAIR	31411	2536
Caenorhabditis elegans	WormBase	11808	718
Rattus norvegicus	RGD	4167	3143
Oryza sativa	Gramene	34377	2300
Danio rerio	ZFIN	4736	394
Dictyostelium discoideum	DictyBase	5440	221
Candida albicans	CGD	677	491

example, in Chapter 7 we will show an example of a gene expression data set that took experts months to analyze manually; text-based algorithms obtained similar results in minutes. Simple methods to summarize the results of a sequence search algorithm using text reduces the amount of time it may take to comprehend the significance of those similar sequences from about an hour to minutes.

One of the goals of writing this book is to help popularize these methods that have the potential of unraveling new biology, and at the very least will expedite exploration greatly. In each chapter we are careful to provide detailed evaluations of the major methods and present practical examples when possible.

1.5 Guide to this book

The aim of this book is to introduce the interested reader to computational strategies to include the scientific text in the analysis of genomics data. The work presented in this book represents the state of the art in the field. Many of the chapters address specific challenges in text mining and bioinformatics, including those in gene function assignment, verifying protein–protein interactions, gene expression analysis, and sequence analysis. These chapters describe active research areas, and some of these chapters introduce novel research.

The different chapters approach text at different levels of granularity. The first chapter discusses entire corpora of documents. Chapters 3–5 use subgroups of text; in these chapters text from multiple documents referring to the same gene are combined. Chapters 6–8 use article abstracts individually. Chapters 9 and 10 looks at individual sentences and words.

Chapter 2 provides a brief review of biology and some key experimental methods. It also reviews some of the important algorithms for the analysis of genomics data and presents a brief review of probability and statistics. It is intended as a primer on bioinformatics and acts as background for the remainder of the book. In Chapter 3 we introduce a simple text representation: the word vector. We show how word vectors can be defined for genes and how it can be a simple but useful tool in bioinformatics analysis. In Chapter 4 we show how sequence analysis can be refined with textual information using gene word vectors in addition to sequences. Chapter 5 focuses on gene expression analysis; we demonstrate how the scientific literature can be used to distinguish true positives from false positives and assess experiment quality. The methods introduced in Chapter 5 can be applied to the results of any assay that assigns a single value to a large number of genes. In Chapter 6 we introduce strategies to assess the extent to which a group of genes contain related genes using only articles about those genes. Since most genomics analyses produce groups of genes, this approach can be easily used in many types of analysis. Chapter 7 focuses on application of the strategies introduced in Chapter 6 to analyze large gene expression data sets in several different organisms. Chapter 8 introduces machine learning on scientific text and demonstrates how it can be useful in gene annotation. In Chapters 9 and 10 we talk about learning relationships between proteins and using the text to learn biological networks between genes and proteins. Chapter 9 introduces methods to find gene names in text. Chapter 10 introduces methods to identify and delineate relationships between genes and proteins in text. We discuss how well text-based methods can be used to verify experimentally predicted interactions. We also explore the possibility of mining the text to create networks of genes.

References

Ashburner, M., C. A. Ball, et al. (2000). "Gene Ontology: tool for the unification of biology. The Gene Ontology Consortium." *Nat. Genet.* **25**(1): 25–9.

Blake, J. A., J. E. Richardson, et al. (2002). "The Mouse Genome Database (MGD): the model organism database for the laboratory mouse." *Nucleic Acids Res.* **30**(1): 113–5.

Blaschke, C., M. A. Andrade, et al. (1999). "Automatic extraction of biological information from scientific text: protein–protein interactions." *Proc. Int. Conf. Intell. Syst. Mol. Biol.* 2(1): 60–7.

Boeckmann, B., A. Bairoch, et al. (2003). "The SWISS-PROT protein knowledgebase and its supplement TrEMBL in 2003." *Nucleic Acids Res.* **31**(1): 365–70.

Brown, P. O. and D. Botstein (1999). "Exploring the new world of the genome with DNA microarrays." *Nat. Genet.* **21**(1 Suppl): 33–7.

Chang, J. T., S. Raychaudhuri, et al. (2001). "Including biological literature improves homology search." *Pac. Symp. Biocomput.* **14**(5): 374–83.

Cherry, J. M., C. Adler, et al. (1998). "SGD: Saccharomyces Genome Database." *Nucleic Acids Res.* **26**(1): 73–9.

Craven, M. and J. Kumlien (1999). "Constructing biological knowledge bases by extracting information from text sources." *Proc. Int. Conf. Intell. Syst. Mol. Biol.* **10**(1): 77–86.

Donaldson, I., J. Martin, et al. (2003). "PreBIND and Textomy–mining the biomedical literature for protein-protein interactions using a support vector machine." *BMC Bioinformatics* **4**(1): 11.

FlyBase (2002). "The FlyBase database of the Drosophila genome projects and community literature." *Nucleic Acids Res.* **30**(1): 106–8.

Fukuda, K., A. Tamura, et al. (1998). "Toward information extraction: identifying protein names from biological papers." *Pac. Symp. Biocomput.* **61**(5): 707–18.

Humphreys, K., G. Demetriou, et al. (2000). "Two applications of information extraction to biological science journal articles: enzyme interactions and protein structures." *Pac. Symp. Biocomput.* **6**(4): 505–16.

Hutchinson, D. (1998). Medline for health professionals: how to search PubMed on the Internet. Sacramento, New Wind.

Jenssen, T. K., A. Laegreid, et al. (2001). "A literature network of human genes for high-throughput analysis of gene expression." *Nat. Genet.* **28**(1): 21–8.

Kwok, P. Y. and X. Chen (2003). "Detection of single nucleotide polymorphisms." *Curr. Issues Mol. Biol.* **5**(2): 43–60.

Lesk, A. M. (2002). *Introduction to Bioinformatics.* Oxford, Oxford University Press.

Liu, F., T. K. Jenssen, et al. (2004). "FigSearch: a figure legend indexing and classification system." *Bioinformatics.* **20**(16): 2880–2.

MacCallum, R. M., L. A. Kelley, et al. (2000). "SAWTED: structure assignment with text description–enhanced detection of remote homologues with automated SWISS-PROT annotation comparisons." *Bioinformatics.* **16**(2): 125–9.

Manning, C. M. and H. Schutze (1999). *Foundations of Statistical Natural Language Processing.* Cambridge, The MIT Press.

Marcotte, E. M., I. Xenarios, et al. (2001). "Mining literature for protein–protein interactions." *Bioinformatics.* **17**(4): 359–63.

Mewes, H. W., D. Frishman, et al. (2000). "MIPS: a database for genomes and protein sequences." *Nucleic Acids Res.* **28**(1): 37–40.

Ng, S. K. and M. Wong (1999). "Toward Routine Automatic Pathway Discovery from On-line Scientific Text Abstracts." *Genome Inform Ser Workshop Genome Inform.* **10**(8): 104–112.

Ohta, Y., Y. Yamamoto, et al. (1997). "Automatic construction of knowledge base from biological papers." *Proc. Int. Conf. Intell. Syst. Mol. Biol.* **5**: 218–25.

Ono, T., H. Hishigaki, et al. (2001). "Automated extraction of information on protein–protein interactions from the biological literature." *Bioinformatics.* **17**(2): 155–61.

Proux, D., F. Rechenmann, et al. (2000). "A pragmatic information extraction strategy for gathering data on genetic interactions." *Proc. Int. Conf. Intell. Syst. Mol. Biol.* **8**(26): 279–85.

Pruitt, K. D. and D. R. Maglott (2001). "RefSeq and LocusLink: NCBI gene-centered resources." *Nucleic Acids Res.* **29**(1): 137–40.

Raychaudhuri, S., J. T. Chang, et al. (2003). "The computational analysis of scientific literature to define and recognize gene expression clusters." *Nucleic Acids Res.* **31**(15): 4553–60.

Raychaudhuri, S., J. T. Chang, et al. (2002). "Associating genes with gene ontology codes using a maximum entropy analysis of biomedical literature." *Genome Res.* **12**(1): 203–14.

Roberts, R. J. (2001). "PubMed Central: The GenBank of the published literature." *Proc. Natl. Acad. Sci. USA* **98**(2): 381–2.

Sekimizu, T., H. S. Park, et al. (1998). "Identifying the Interaction between Genes and Gene Products Based on Frequently Seen Verbs in Medline Abstracts." *Genome Inform Ser Workshop Genome Inform.* **9**: 62–71.

Sharff, A. and H. Jhoti (2003). "High-throughput crystallography to enhance drug discovery." *Curr. Opin. Chem. Biol.* **7**(3): 340–5.

Shatkay, H., S. Edwards, et al. (2000). "Genes, themes and microarrays: using information retrieval for large-scale gene analysis." *Proc. Int. Conf. Intell. Syst. Mol. Biol.* **8**(10): 317–28.

Shatkay, H. and R. Feldman (2003). "Mining the biomedical literature in the genomic era: an overview." *J. Comput. Biol.* **10**(6): 821–55.

Stein, L., P. Sternberg, et al. (2001). "WormBase: network access to the genome and biology of Caenorhabditis elegans." *Nucleic Acids Res.* **29**(1): 82–6.

Stephens, M., M. Palakal, et al. (2001). "Detecting gene relations from Medline abstracts." *Pac. Symp. Biocomput.* **52**(3): 483–95.

Tamames, J., C. Ouzounis, et al. (1998). "EUCLID: automatic classification of proteins in functional classes by their database annotations." *Bioinformatics.* **14**(6): 542–3.

Thomas, J., D. Milward, et al. (2000). "Automatic extraction of protein interactions from scientific abstracts." *Pac. Symp. Biocomput*: 541–52.

Venter, J. C., M. D. Adams, et al. (2001). "The sequence of the human genome." *Science* **291**(5507): 1304–51.

Wong, L. (2001). "PIES, a protein interaction extraction system." *Pac. Symp. Biocomput.* **233**(1473): 520–31.

Yandell, M. D. and W. H. Majoros (2002). "Genomics and natural language processing." *Nat. Rev. Genet.* **3**(8): 601–10.

Zhu, H., M. Bilgin, et al. (2003). "Proteomics." *Ann. Rev. Biochem.* **72**: 783–812.

2 Functional genomics

The overarching purpose of this chapter is to introduce the reader to some of the essential elements of biology, genomics, and bioinformatics. It is by no means a comprehensive description of these fields, but rather the bare minimum that will be necessary to understand the remainder of the book.

In the first section we introduce the primary biological molecules: nucleic acids and proteins. We discuss genetic information flow in living beings and how genetic material in DNA is translated into functional proteins. In the second section we present a short primer on probability theory; we review some of the basic concepts. In the third section we describe how biological sequences are obtained and the common strategies employed to analyze them. In the fourth section, we describe the methods used to collect high throughput gene expression data. We also review the popular methods used to analyze gene expression data.

There are many other important areas of functional genomics that we do not address at all in this chapter. New experimental and analytical methods are constantly emerging. For the sake of brevity we focused our discussion on the areas that are most applicable to the remainder of the book. But, we note that many of the analytical methods presented here can be applied widely and without great difficulty to other data types than the ones they have been presented with.

2.1 Some molecular biology

Here we present a focused review of molecular biology designed to give the reader a sufficient background to comprehend the remainder of the book. A thorough discussion is beyond the scope of this book and the interested reader is referred to other textbooks (Alberts, Bray et al. 1994; Stryer 1995; Nelson, Lehninger et al. 2000).

2.1.1 Central dogma of molecular biology

The central dogma of molecular biology is a paradigm of information flow in living organisms (see Plate 2.1). Information is stored in the genomic deoxyriboculeic acid (DNA). DNA polymerase, a protein that synthesizes DNA, can replicate DNA so that it can be passed on to progeny after cell division. During transcription, RNA polymerase, a protein that synthesizes RNA, uses the information from genes contained in the DNA sequence to produce messenger ribonucleic acid (mRNA). During translation, the ribosomal complex then uses mRNA as a template to synthesize proteins. Proteins are involved in most biological processes and have a wide range of functions, including enzymatic activity, transport, storage, and providing structural integrity to a cell. In general it is the presence and activity of the proteins that make a given cell unique and that permit a cell to react to physiological circumstances or stresses.

2.1.2 Deoxyribonucleic acid

Deoxyribonucleic acid (DNA) molecules are the building blocks of life. They contain the genetic material that is passed on through generations of a species; they contain the blueprints for all of the proteins in an organism. DNA is a long threadlike molecule composed by a linear chain of deoxyribonucleotide bases. There is great variability in the amount of DNA in the genomes of different organisms (see Table 2.1).

DNA is composed of deoxyribonucleotides, which is constituted from a deoxyribose, a sugar moiety, one or more phosphate groups, and a nucleotide base (see Figure 2.1). Deoxyribose is similar to the sugar ribose, except it is lacking an oxygen molecule at the 2′ carbon. There are four possible nucleotide bases that can be employed in DNA: adenosine (A) and guanine (G) are purine bases, while cytosine (C) and thymine (T) are pyrimidine bases (see Figure 2.2).

Table 2.1 *Genome size for different species.*

Organism	Description	Bases	Genes
Epstein barr virus	Virus	170,000	80
Mycobacterium tuberculosis	Bacteria	4,400,000	3959
E. Coli	Bacteria	4,600,000	4377
Plasmodium falciparum	Malaria parasite	23,000,000	5268
Drosophila melanogaster	Fruit Fly	123,000,000	13,472
Humans		3,000,000,000	30,000

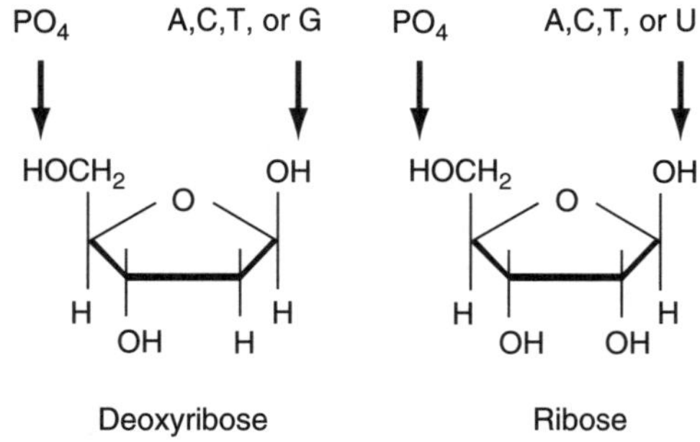

Figure 2.1 *Deoxyribose and ribose.* Deoxyribose and ribose are sugar moieties that are the key structural component of DNA and RNA bases respectively. In this figure the chemical structures of these sugars are illustrated. Phosphate groups and bases bind the sugar moieties at the indicated position to form deoxyribonucleotides and ribonucleotides.

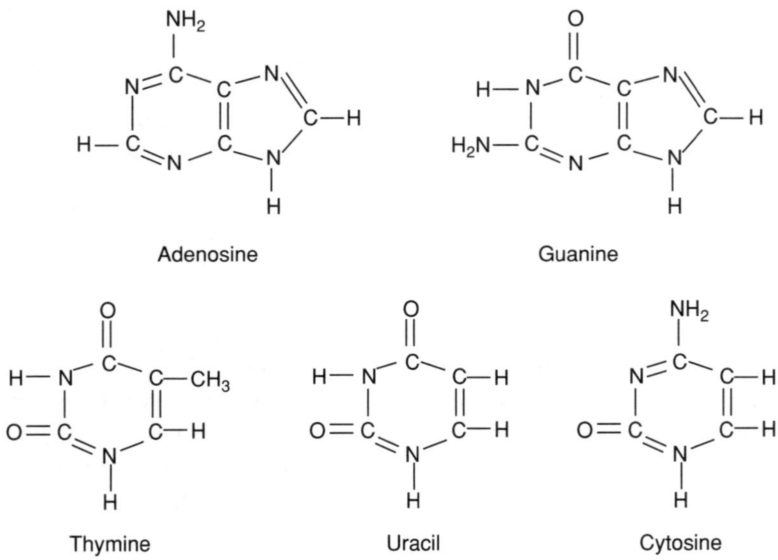

Figure 2.2 *Nucleotide bases.* The nucleotide bases are depicted here. The purines are illustrated above; they are adenosine and guanine, while the pyrimidines are depicted below; they are thymine, uracil, and cytosine. Adenosine can hydrogen bond with thymine or uracil. Guanine and cytosine can hydrogen bond as well. Adenosine, guanine, thymine, and cytosine are commonly found in DNA. In RNA, uracil is found instead of thymine.

Deoxyribonucleotides can be linked together through their phosphate groups to form DNA polymer chains as depicted in Figure 2.3. The invariant DNA backbone consists of the deoxyribose sugar and phosphate groups in an alternating repetitive fashion. The bases, however, are the variable element in DNA. The information contained in DNA is encoded within the sequence of bases. Conventionally, DNA sequences are written to indicate the bases in order from the 5′ carbon end to the 3′ carbon end with the letters A, C, T, and G.

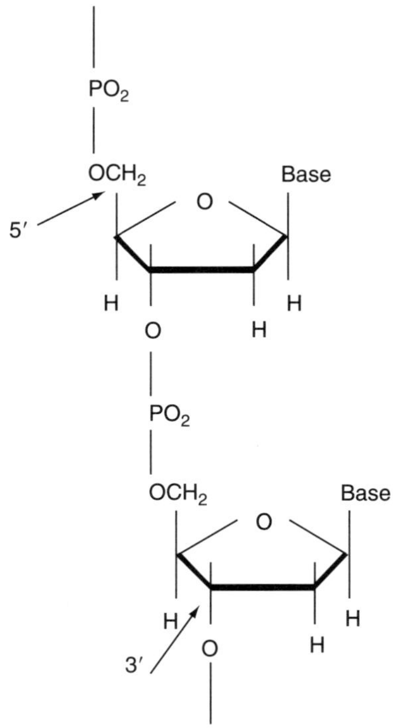

Figure 2.3 *The phosphodiester bond.* The phosphodiester bond links nucleotides together. Depicted here are two nucleotides in a DNA polymer. The 5′ nucleotide is at the top, and the 3′ nucleotide is at the bottom. The phosphate group from the 3′ nucleotide binds the 3′ carbon of the 5′ nucleotide to form the phosphodiester bond. DNA molecules are composed of millions of bases strung together in this fashion.

The native three-dimensional structure of DNA is a double stranded helix about 20 angstroms in diameter, in which the bases from both helices are in the interior of the helix, and the backbone is on the exterior. The structure of DNA is contingent on pairing between the bases. Through hydrogen bonding, adenosine is able to pair with thiamine, and guanine can pair with cytosine. But, the pairings are in the opposite direction. Consequently DNA exists as two chains complementing each other running antiparallel (see Figure 2.4).

2.1.3 Ribonucleic acid

Ribonucleic acid (RNA) is similar to DNA in that it is a nucleic acid polymer. However, instead of deoxyribonucleotides, RNA is composed of

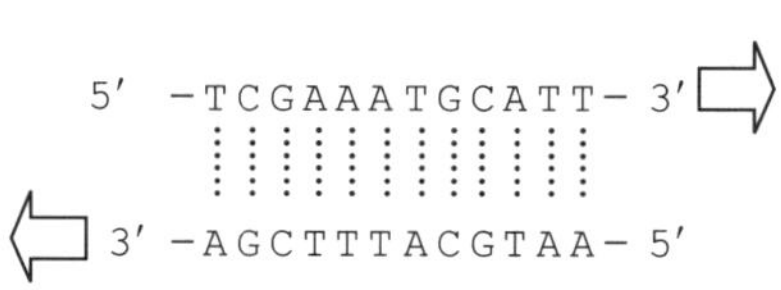

Figure 2.4 *DNA base pairing.* DNA pairs with its reverse-complement strand in an anti-parallel fashion. So a DNA strand that runs from 5′ to 3′ will pair with a strand oriented in the opposite direction. For hydrogen bonding to be satisfied A and T nucleotides must line up, as must G and C nucleotides. The two strands wrap together and form a double helix where the nucleotides are on the interior and the phosphate backbone is on the exterior.

ribonucleotides; the key difference is that ribose serves as the sugar moiety instead (see Figure 2.1). RNA has a ribose and phosphate backbone. The bases are the same except for thiamine; it is substituted by uracil (U) as a pyrimidine base pair. Uracil has similar hydrogen bonding properties to thymine and can also pair with adenosine. Since the RNA bases are similar to DNA bases, RNA can hybridize to DNA. Unlike DNA, RNA is rarely found in double stranded forms and can be found having many different complex structures (see Plate 2.2); it can fold back on itself and self-hybridize, creating hairpin loops (see Figure 2.5). Base pairing is sometimes imperfect in RNA.

RNA polymerase transcribes a DNA sequence into RNA by using DNA as a template during RNA synthesis. The DNA templates have special short sequences called promoter sites that indicate locations on the DNA strand that RNA polymerase can bind and start transcription. RNA polymerase synthesizes RNA in the 5′ to 3′ direction by using DNA as a base pair template; the bases of the synthesized RNA are the reverse complement of the template DNA strand. Consequently RNA bases match those of the reverse complement strand of the DNA template perfectly, except that thymine bases are replaced by uracil.

Functionally there are three major categories of RNA. Ribosomal RNAs, or rRNA, are structural components of the ribosomal complex and are involved in protein synthesis. Transfer RNAs, or tRNA, are small molecules consisting of 70–90 base pairs that have a characteristic three-dimensional structure (see plate 2.2) (Shi and Moore 2000). They carry individual amino acids that are polymerized during protein synthesis. Finally, messenger RNA, mRNA, is the carrier of genetic information from DNA and acts as a template for protein synthesis.

After synthesis, mRNA may be modified. In eukaryotes a cap sequence is added on the 5′ end, and after cleaving extraneous nucleotides at the 3′ end, a long poly-A tail consisting of about 250 adenosine nucleotides is added. In addition regions of mRNA called introns are spliced out, leaving other regions known as exons that will be translated into proteins (see Figure 2.6).

The mRNA sequence can then be used for protein synthesis. Each series of three nucleotides in the mRNA is called a codon. Each codon corresponds to either a specific amino acid, a signal to start translation, or stop.

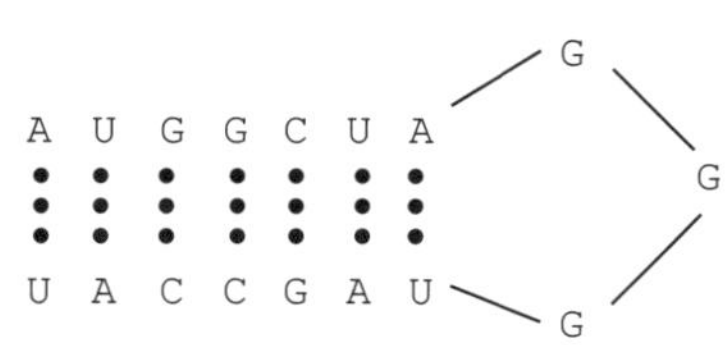

Figure 2.5 *RNA hairpin loop*. Here we illustrate a hypothetical mRNA hairpin loop. The mRNA literally folds in on itself and self-hybridizes. In the process a tight loop is formed.

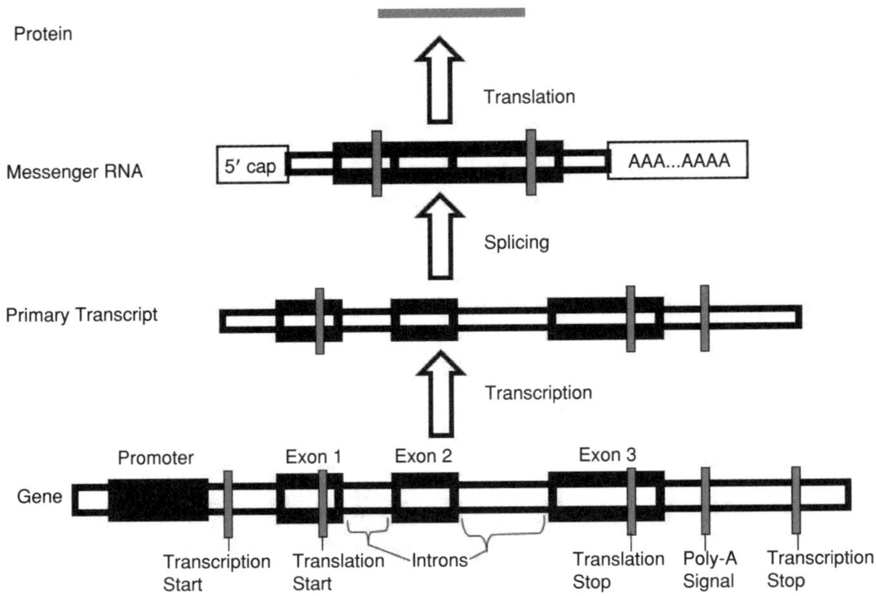

Figure 2.6 *From gene sequence to protein.* The gene is physically embedded in the genomic DNA sequence. The lowest bar represents the DNA gene sequence. Black and grey boxes indicate specific sequences in the gene with functional significance. Genes have promoter sites that bind the transcription machinery. DNA transcription starts at the transcription start site and extends to the transcription stop site. A primary transcript, which is the precursor to mRNA, is produced. A 5′ cap is appended. The transcript is cleaved at the poly-A signal site and a poly-A tail is appended. Introns are cut out of the transcript, and the messenger RNA is produced. The messenger RNA consists of the 5′ and 3′ untranslated regions, the exons, and the appended poly-A tail and 5′ cap. The mRNA is then translated into protein. Translation occurs between the translation start and stop sites.

The genetic code specifies the corresponding amino acid for each codon (see Table 2.2). Codons are recognized by specific tRNA molecules that carry the appropriate amino acid that corresponds to the codon in the genetic code. Those amino acids are bound together to synthesize a polypeptide.

2.1.4 Genes

Genes are the functional and physical units of heredity that are passed from parent to offspring. In the modern era genes are increasingly thought of as a segment of DNA sequence that corresponds to a particular protein. Classically, scientists have studied genes by observing phenotypes, or observable traits, in organisms and how they were transmitted to their offspring. The observable traits are accounted for by gene sequence variations. At the molecular level, these variations usually confer differences in the protein product's structure or production.

Gene sequences have complex structures including a promoter region where the RNA transcription machinery binds, a transcription initiation

Table 2.2 *The genetic code.* Here we list each of the 64 possible codons. Each codon is three nucleotides in the messenger RNA or DNA. Each codon corresponds to a particular amino acid. For example TTT corresponds to phenylalanine and GTT corresponds to valine. Note that some codons specifically indicate the end of the protein. Many amino acids have multiple codons and the third position in the codon is usually the most degenrate.

	Position 2				
Position 1	T	C	A	G	Position 3
	Phe	Ser	Tyr	Cys	T
T	Phe	Ser	Tyr	Cys	C
	Leu	Ser	END	END	A
	Leu	Ser	END	Trp	G
	Leu	Pro	His	Arg	T
C	Leu	Pro	His	Arg	C
	Leu	Pro	Gln	Arg	A
	Leu	Pro	Gln	Arg	G
	Ile	Thr	Asn	Ser	T
A	Ile	Thr	Asn	Ser	C
	Ile	Thr	Lys	Arg	A
	Met	Thr	Lys	Arg	G
	Val	Ala	Asp	Gly	T
G	Val	Ala	Asp	Gly	C
	Val	Ala	Glu	Gly	A
	Val	Ala	Glu	Gly	G

site where RNA synthesis begins, and a transcription stop site where RNA synthesis is terminated. A translation start site indicates the position at which protein synthesis begins; a translation stop site indicates where protein synthesis ends. The 5′ untranslated region is between the transcription initiation site but before the translation start site; it codes DNA that will become part of the RNA transcript, but will not affect protein synthesis. Similarly the 3′ untranslated region is after the translation stop site but before the transcription stop site. Eukaryotic genes have a polyadenylation signal which specifies the position at which the mRNA is cleaved and a poly-A tail is appended. Eukaryotic genes also have introns that are spliced out of transcript before protein translation occurs. Figure 2.6 provides a schematic of a gene structure.

Gene expression, or mRNA transcription, is partially regulated locally at the promoter regions of a gene. Promoter regions are located upstream of the transcribed region of a gene on the same DNA strand. They bind transcription factor proteins that give RNA polymerase access to the gene and permit RNA transcription. These sequences can bind other activating and repressing proteins that affect gene expression depending on the physiologic condition that the cell is in. Enhancers are other distant regulatory sequences found in eukaryotes that can be located thousands of base pairs away from the gene and can be located on either DNA strand. Stimulatory proteins can bind enhancers and affect the expression of distant genes.

2.1.5 Proteins

Proteins are the workhorses of an organism. They are intimately involved in all biological processes. Enzymes catalyze biological reactions, including those that are involved in metabolism and catabolism. Transport proteins facilitate the movement of small and large molecules through membranes. Transcription factors are special proteins that regulate the production of mRNA for specific genes. Chaperones help other proteins fold into the right shape. Polymerase proteins facilitate DNA replication, and RNA synthesis. Structural proteins give cells shape and stability.

Proteins are linear chains of amino acids. An amino acid is composed on a central carbon atom attached to an amino group, a carboxyl group, and a distinctive side chain (see Figure 2.7a). There are 20 different amino acids, each with unique biochemical properties (see Table 2.3). They are linked together by a strong peptide bond between the amide and carboxyl groups of the amino acids (see Figure 2.7b). Amino acid sequences are written from the amino (N) to the carboxyl (C) end. Since some amino acids are more similar to each other than others, it has been observed when comparing similar but evolutionarily related proteins that certain pairs of amino acids are statistically more likely to substitute for each other.

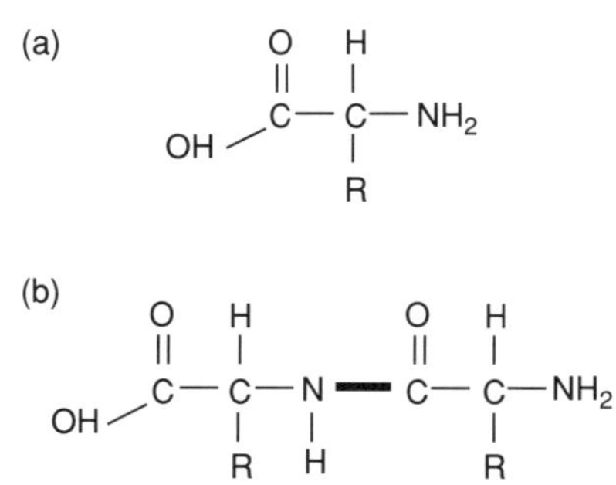

Figure 2.7 (a). *Basic amino acid structure.* The basic amino acid structure is indicated above. Each amino acid has a central carbon. It is attached to a hydrogen atom, a carboxyl (COOH) group, and an amino group. It is also attached to a variable R group. There are 20 different possibilities for the R groups, each resulting in a different amino acid with different biochemical properties. (b) *The peptide bond.* This is a schematic illustration of the peptide bond. A hydrogen atom from the amino group of one amino acid and a hydroxyl group from the other amino acid are removed for a net loss of a single water molecule. The nitrogen from the amino group and the carbon from the carboxyl group are bound together. This process is repeated many times for polypeptides. The C-terminal end of this short dipeptide is on the left; there is a free carboxyl group. The N-terminal end of this peptide is on the right; there is a free amino group.

Table 2.3 *The amino acids.* We list the 20 different amino acids along with both their three- and one-letter abbreviations. In addition we list the chemical structure of their side chain. In the final column we list their salient chemical property.

Amino acid	3-letter symbol	1-letter symbol	Side chain	Chemical properties
Alanine	ala	A	CH^3–	Alipathic
Arginine	arg	R	HN = C(NH^2)–NH–$(CH^2)^3$–	Basic
Asparagine	asn	N	H^2N–CO–CH^2–	Acidic
Aspartic acid	asp	D	HOOC-CH^2–	Acidic
Cysteine	cys	C	HS-CH^2–	Sulfur group
Glutamine	gln	Q	H^2N–CO–$(CH^2)^2$–	Acidic
Glutamic acid	glu	E	HOOC–$(CH^2)^2$–	Acidic
Glycine	gly	G	NH^2–CH^2–COOH	Alipathic
Histidine	his	H	NH–CH = N–CH = C–CH^2–	Basic
Isoleucine	ile	I	CH^3–CH^2–CH(CH^3)–CH(NH^2)–COO	Alipathic
Leucine	leu	L	$(CH^3)^2$–CH–CH^2–	Alipathic
Lysine	lys	K	H^2N–(CH^2)4–	Basic
Methionine	met	M	CH^3–S–$(CH^2)^2$–	Sulfur group
Phenylalanine	phe	F	Ph-CH^2–	Aromatic group
Proline	pro	P	NH–$(CH^2)^3$–CH–COOH	Imino acid
Serine	ser	S	HO–CH^2–	Hydroxyl group
Threonine	thr	T	CH^3–CH(OH)–	Hydroxyl group
Tryptophan	trp	W	Ph–NH–CH = C–CH^2–	Aromatic group
Tyrosine	tyr	Y	HO–p–Ph–CH^2–	Aromatic group
Valine	val	V	$(CH^3)^2$–CH–	Alipathic

The amino acid sequence of a protein, known as a protein's primary structure, determines the protein's three dimensional structure and function. Non-covalent hydrogen bonding between the amino and carboxyl groups from different amino acids allows the linear chain to fold into structures known as alpha helices and beta sheets. Beta sheets are formed when peptide strands line up next to each other in either a parallel or antiparallel fashion (Figure 2.8). On the other hand, alpha helices are formed when peptide strands coil up; amino acids hydrogen bond to other amino acids above and below it in the helix (Plate 2.3). These structures constitute the protein's secondary structure. Protein function is largely dictated by the protein's three-dimensional, or tertiary, structure. A ribbon diagram depicting the backbone of a protein is displayed in Plate 2.4 (Williams, Zeelen et al. 1999). Note that the secondary structure is an evident feature in this structure.

Proteins are capable of forming complexes with other proteins and also interacting transiently with them. Proteins can often bind DNA or RNA. For example, ribosomal proteins involved in protein synthesis may bind

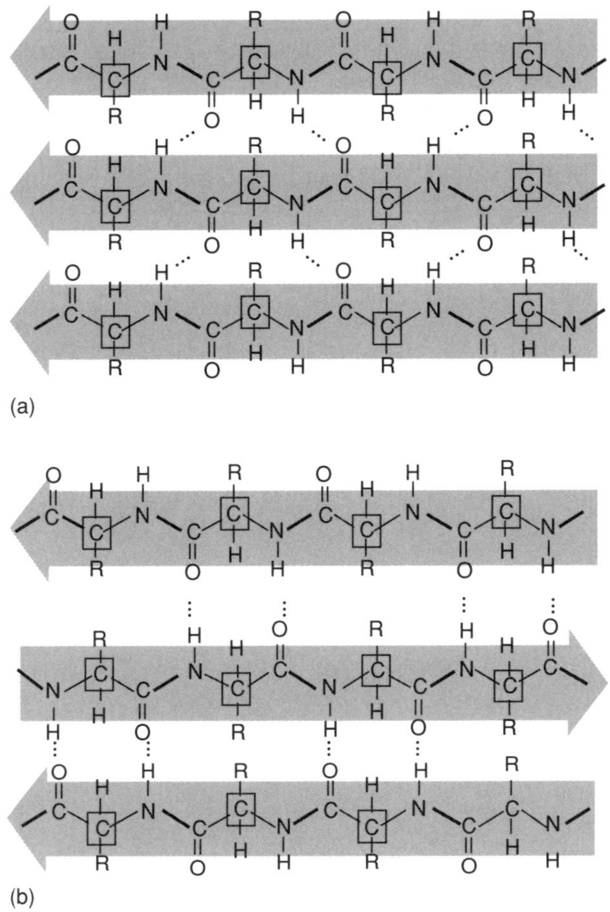

Figure 2.8 *Hydrogen bonding in beta sheets.* (a) Parallel beta strands. The hydrogen bonding between parallel peptide strands is depicted in this figure with dotted lines. The arrow indicates the N to C direction of the peptide. The boxed C indicates the amino acid alpha carbon. Hydrogen bonding is between the amino hydrogen and the carboxyl oxygen. (b) Antiparallel beta strands. Hydrogen bonding between antiparallel beta sheets is similarly depicted.

mRNA, whereas transcription factors that regulate gene expression would bind specific DNA sites.

2.1.6 Biological function

One of the goals of functional genomics and its approaches is to understand what the function of each of the genes and their protein products is. By function, we mean what the role of the gene or protein is in an organism. To this end, many modern high throughput functional assays are becoming available. For example yeast-2-hybrid assays can assess what other proteins a single protein binds to. Gene expression microarrays can indicate the conditions under which genes are expressed. Systemic deletion approaches can demonstrate gene function by demonstrating functional deficiencies in an organism when the gene is removed.

But, function is a very protean concept, and it can mean very different things in different contexts. For example, a medical geneticist might note that people lacking a particular copy of a specific gene are at high risk of cancer; she may say that the function of the protein product of the gene is "tumor suppression". On the other hand a cell biologist may ascertain that the protein product of the gene localizes to the nucleus; she may suggest that the protein functions as a "nuclear protein". A biochemist may use sophisticated assays to determine that the protein binds to DNA and may say it is a "DNA binding" protein. A molecular biologist may render the gene dysfunctional in individual cells and see that cell growth arrests, and decide that the gene is a "cell growth" gene. All of these same functions could coexist for the same gene or protein, and all are very relevant to understanding the role of a gene in an organism.

It is important to realize that, as we move forward in this new era of genomics, while high throughput functional assays may give broad functional information about all the genes simultaneously in an organism, the type of functional information we learn can be quite narrow and often ambiguous.

2.2 Probability theory and statistics

In this section we present a brief review of probability theory. Here we focus on the basics that are necessary for this book. For many readers this will be at a very elementary level. Other readers are encouraged to spend time to thoroughly understand this material, as it is the cornerstone for the methods that we will present, and indeed, for much of bioinformatics.

The content of this section is described in the frame box. This section introduces the reader to basic concepts in probability theory, probability distribution functions, concepts in information theory, statistical measures, and performance measures.

2.2.1 Probability

A probability is a number between 0 and 1 that is assigned to a particular observation or event; the value is proportional to our expectation of the likelihood of the event occurring. We say the probability of an event is 0 if it can never happen, and is 1 if it will always happens. The number assigned represents the fraction of times that the event is expected to occur over a large

1) Probability theory
   a) Conditional probability
   b) Probabilistic independence
   c) Bayes' theorom
2) Probability distribution functions
   a) Binomial distribution
   b) Poisson distribution
3) Information theory
   a) Entropy of a distribution
   b) Kullback–Liebler distance
4) Population statistics
   a) Mean
   b) Median
   c) Variance
   d) Z-scores
5) Performance measures
   a) Accuracy
   b) Sensitivity and specificity
   c) Precision and recall

number of trials. We say the probability of an event A is $P(A)$. Frequently, we assume that if an event A has been observed to occur n times out of a large number of N instances then $P(A)$ is equal to n/N. To illustrate using a simple example one can consider a fair coin toss. The likelihood of the coin landing heads side up is equal to that of the coin landing tails side up. If the coin is tossed 100 times, 50 of the observed outcomes will be heads. Thus if we denote the heads outcome as event A, $P(A)$ is found to be 0.5.

We speak of the probability space as the collection of all possible events that can happen. In general the sum of all of the possible observations must be 1; that is, no matter what, at least one of the possible observations must be observed. In the case of the fair coin toss the probability space is comprised of heads or tails, the two possible outcomes that could occur for the event.

2.2.2 Conditional probability

Often, the probability of an observation depends on the occurence of other events. Conditional probability is the probability of an event occurring in the context of another event. For example, consider the probability that a house has been burglarized. That probability might be increased if it is known that a burglar has been spotted in the neighborhood. On the other hand, that probability might be reduced if it is known that the house has an alarm in good working order that has not been set off. So the probability of the event is dependent on other events or other knowledge items that influence our expectation that the event has or has not occurred. Formally we define conditional probability as:

$$P(A|B) = \frac{P(A, B)}{P(B)}$$

Here we say that the probability of event A is being conditioned on B. $P(A|B)$ is the probability of event A given that event B is known to have occurred and $P(A, B)$ is the probability that both event A and event B have occurred. We can rearrange the equation above to read as:

$$P(A, B) = P(A|B)P(B)$$

This is a particularly useful expression. Often times we know the probability of an event, and the probability of a second event given the first. The above equation allows us to convert those probabilities into a probability of both events occurring.

We can extend the above expression in general:

$$P(A_1, A_2 \ldots A_n) = P(A_1)P(A_2|A_1)P(A_3|A_1, A_2)$$
$$P(A_4|A_1, A_2, A_3) \ldots P(A_n|A_1, A_2, \ldots, A_{n-1})$$

2.2.3 Independence

We say two events are independent of each other one when the possibility of one event occurring is not affected in any way by the other event. Formally we define independence between two events A and B:

$$P(A|B) = P(A)$$

So the probability of A is unchanged regardless of the outcome of B. For example the probability of a coin being heads on the second flip is 1/2. Whether the coin is heads or tails on the first flip has no influence on this. On the other hand, the probability that an ace is the second card drawn from a deck of cards is intimately dependent on whether or not an ace was drawn first. If an ace was not drawn, the probability is 4/51. If it were drawn then it is 3/51. So here, we cannot say that the probabilities of the second draw and first draw being aces is independent. The independence assumption can be restated after some rearrangement:

$$P(A, B) = P(A)P(B)$$

In fact, we can extend this relationship to n independent events:

$$P(A_1, A_2 \ldots A_n) = P(A_1)P(A_2)P(A_3)P(A_4) \ldots P(A_n)$$

This is a very useful equation if we know the probabilities of multiple events occurring, and can assume that they are independent of one another. For this reason, the independence assumption is often used in bioinformatics and in text mining, even if it does not apply perfectly, as it provides easy to compute, quick and dirty estimates of an observation. For example, the probability of an amino acid sequence can be estimated as the multiplicative probabilities of each of its amino acids. While this is far from accurate, it offers a simple starting point.

2.2.4 Bayes' theorem

Bayes' theorem allows us calculate the conditional probability of an event using probabilities of the conditions given the event. For example, we may know the probability that lung cancer patients have a smoking history. We can use that information to calculate the probability of lung cancer in smokers. We use the definition of conditional probability:

$$P(A|B) = \frac{P(A, B)}{P(B)} = \frac{P(B|A)P(A)}{P(B)}$$

If all of the possible observations for A are A_i for all i, then we can make the substitution:

$$P(B) = \sum_i P(A_i, B)$$

and further rearrangement gives:

$$P(A|B) = \frac{P(B|A)P(A)}{\sum_i P(B|A_i)P(A_i)}$$

Bayes' theorem has many practical applications in bioinformatics. It can be used to update our probability of an event given addition knowledge. For example, we may wish to calculate the probability of a protein sequence having a biological function A, given that an assay B is positive.

Assume that we know the following parameters: (1) the prior probability of the biological function A, $P(A)$, (2) the probability of assay B being positive if the assayed protein has function A, $P(B|A)$, and (3) the probability of assay B being positive if the protein does not have function A, $\mathrm{P}(B|\bar{A})$. We can use the above formula to determine the probability that this protein does in fact have function A in the context of the new assay information. In this case the formula reduces to:

$$P(A|B) = \frac{P(B|A)P(A)}{P(B|A)P(A) + P(B|\bar{A})(1 - P(A))}$$

2.2.5 Probability distribution functions

In many cases there are many possible outcomes that might occur for a given event. In these cases we can define probability distribution functions (pdfs). These functions can either be continuous or discrete depending on the nature of the probability space they correspond to. They assign numerical probability values to each of the possible observations. They must have positive values for each of the possible observations. In addition the total probability over all possible observations must be one:

$$\sum_{x \in \chi} f(x) = 1$$

$$f(x) > 1 \quad x \in \chi$$

where χ is the space of all possible observations, and f is a discrete probability distribution function. If f is a continuous probability distribution we require that the integral of f over the space of possible observations is one:

$$\int_{x \in \chi} f(x) = 1$$

In Figure 2.9 we have displayed some common probability distribution functions.

As an example of a probability distribution consider the example of N binary trials of an event that occurs with probability p. What is the probability that n of those trials is positive? The number of possible ways that n out of N trials can be positive is

$$\binom{N}{n} = \frac{N!}{n!(N-n)!}$$

The probability for each of these trials is:

$$p^n(1-p)^{N-n}$$

So the probability of n out of N trials being positive is:

$$P(n) = \binom{N}{n} p^n(1-p)^{N-n}$$

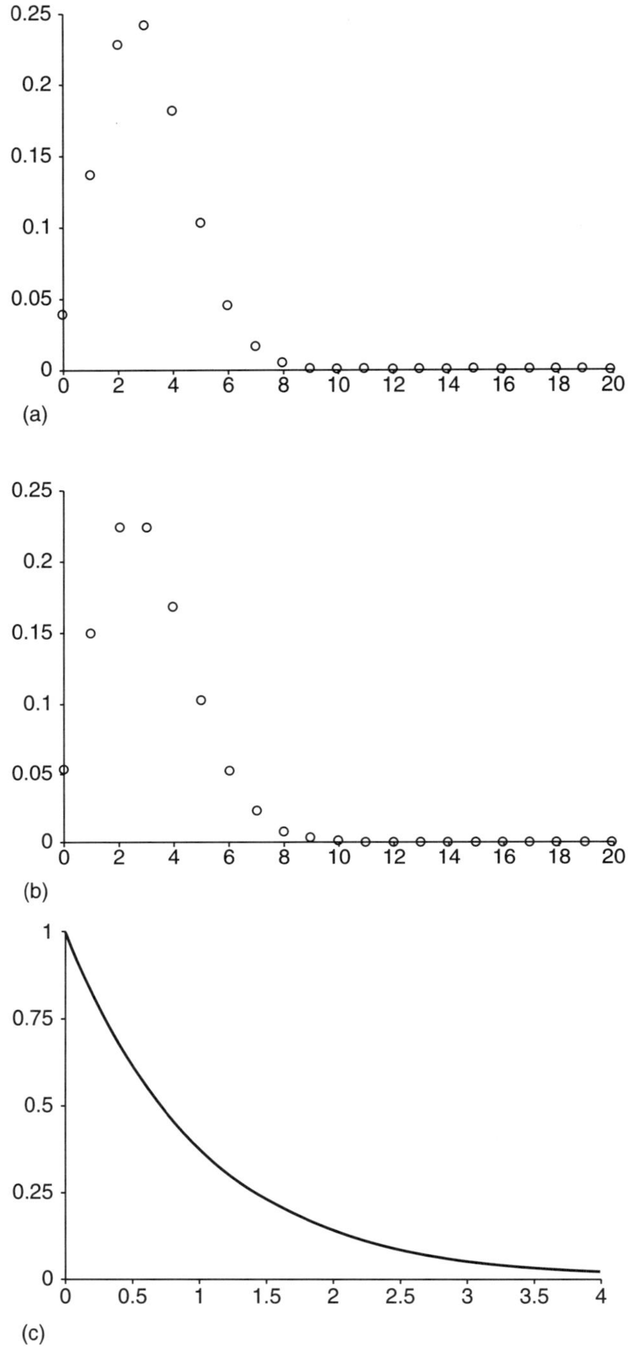

Figure 2.9 *Different probability distribution functions.* (a) *Binomial distribution.* A discrete distribution representing the probability for *n* possible events, where the probability of each event is 0.15, and 20 trials are attempted. (b) *Poisson distribution.* A discrete distribution that approximates the binomial distribution. (c) *Exponential distribution.* One example of a continuous distribution.

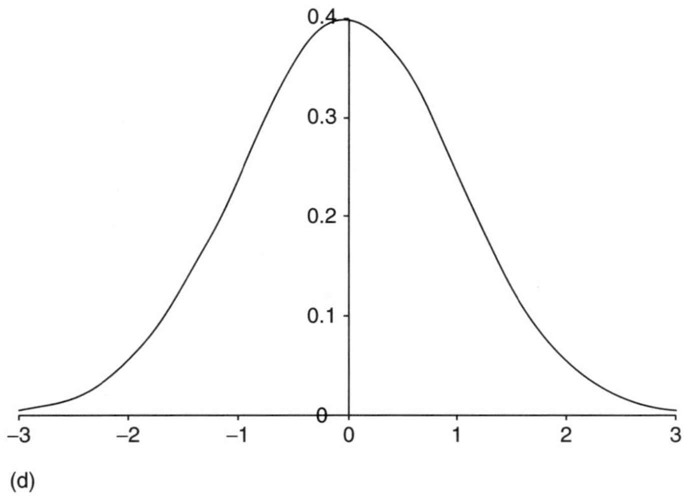

Figure 2.9 (*Continued*) (d)*The normal distribution.* A second example of a continuous distribution.

This discrete probability distribution function is known as the binomial distribution (see Figure 2.9a). The Poisson distribution can be used to approximate this distribution if Np is small:

$$P(n) = \frac{\lambda^n}{n!} e^{-\lambda}$$

where λ is Np. This is illustrated in Figure 2.9(b).

2.2.6 Information theory

This book uses some of the basic concepts from information theory in some of the chapters. The development of information theory stems from transmitting coded messages, but has found wide application in many areas.

Suppose we are trying to transmit information about a sequence of events using binary bits. To create as short a message length as possible, one might select a code so that events that are expected to be more frequent are transmitted with shorter sequence of bits, and rare events can utilize longer sequences of bits. In fact, the code length that will result in the shortest possible message length is

$$-\log_2(p_i)$$

where p_i is the probability of the event i that is being coded. If we have optimally coded a sequence of events that occur with probability p we can calculate the average message length:

$$I = -\sum_i p_i \log_2 (p_i)$$

The above entity is also known as the entropy of the distribution p. If p is the uniform distribution where all circumstances have equal probability, the entropy is the largest possible. In this case we have no cause to expect one event more than another, and we have no prior prejudices. In this case the entropy is maximal. The other extreme of course is if the probability of one of the events is 1, while the probability of the other events is 0, then the entropy of that distribution is zero. In this case we always know exactly what the outcome will be, and there is no reason to expect anything else. Entropy is therefore a measure of the information contained in a distribution.

One way to compare two different distributions is to consider what would happen if we coded events with an incorrect theoretical distribution q, while the actual probability of events was p. The Kullback–Liebler (KL) distance is the average difference in the number of bits in the message and the optimal number of bits. For a given event i this difference in message length is:

$$\log_2 (p_i) - \log_2 (q_i)$$

Averaged over all messages:

$$\mathrm{KL}(p\|q) = \sum_i p_i(\log_2 (p_i) - \log_2 (q_i)) = \sum_i p_i \log_2 \left(\frac{p_i}{q_i}\right)$$

The KL distance is a very helpful means of quantifying the difference between a theoretical and practical distribution.

2.2.7 Population statistics

There are several key statistical parameters that are commonly used to describe a collection, or population, of data. The *mean* and the *median* are measures of the central tendency of the data. The *standard deviation* and *variance* are measures of the spread of the data. These parameters can be calculated for a collection of data and also for a probability distribution function.

The mean, $\bar{x}$, for a collection n data points is calculated simply:

$$\bar{x} = \frac{1}{n}\sum_i x_i$$

where x_i represents the i-th data point. The major limitation of the mean as a measure of central tendency is that it is influenced markedly by extreme values.

So, often, the median is a more valuable measure. When the data points are sorted in increasing order, the median is the value that is in the middle. Half of the data points should be greater in value and half should be less.

For a distribution function, the mean can be calculated as well:

$$\bar{x} = \sum_{x_i \in \chi} x_i f(x_i)$$

For continuous distribution functions the summation is replaced by an integral. For a distribution function, the median is the value that half the probability mass is less than, and half the probability mass is greater than.

The variance, σ^2, can be calculated for a collection of points:

$$\sigma^2 = \frac{1}{n} \sum_i (x_i - \bar{x})^2$$

The variance is a measure of the spread of the data. The standard deviation, σ, is the square root of the variance. If all of the data points are centered right at the mean, the variance is zero. As the points are spread further out, the variance increases. For a probability distribution function:

$$\sigma^2 = \sum_{x_i \in \chi} f(x_i)(x_i - \bar{x})^2$$

For a continuous distribution, the summation is replaced by an integral.

One standardized measure of the "extremeness" of a given data point in the context of a collection of data is the *z-score*. The z-score is a measure of the number of standard deviations that a particular value is away from the mean. We can calculate a z-score as

$$z = \frac{x - \bar{x}}{\sigma}$$

If the data are normally distributed (see Figure 2.9d), then only about 32% of the data has z-scores greater than one or less than negative one; only about 5% of the data has z-scores greater than two or less than negative two.

2.2.8 Measuring performance

As a final issue we introduce some statistical measures that are commonly used to assess the performance of algorithms. Suppose that we have an algorithm that makes binary predictions on a collection of data. The algorithm can make a positive or negative prediction. Some of those predictions might be true, while others might be false. The results can be divided into true positives (TP), false positives (FP), true negatives (TN), and false negatives (FN) – see Figure 2.10. There are a number of key performance measures that we will use repeatedly in this book that we will define here. Sensitivity and specificity are often used to measure the success of machine learning methods. The sensitivity is the fraction of the cases that the algorithm should have called positive that were correctly called. In other words:

$$\text{Sensitivity} = \text{TP}/(\text{TP} + \text{FN})$$

A complementary measure is the specificity. The specificity is the fraction of the cases that the algorithm should have called negative that were correctly called. In other words:

$$\text{Specificity} = \text{TN}/(\text{TN} + \text{FP})$$

The most sensitive approach is to call every case as positive, while the most specific approach is to call every case as negative. Accuracy is the fraction of cases called correctly:

$$\text{Accuracy} = (\text{TP} + \text{TN})/(\text{TP} + \text{TN} + \text{FP} + \text{FN})$$

		Truth	
		Positive	Negative
Prediction	Positive	True Positives	False Positives
	Negative	False Negatives	True Negatives

Figure 2.10 *Prediction results.* The results of any binary predictive algorithm can be organized as above. The class predictions can be divided into the upper and lower rows as positive and negative predictions. Similarly, a gold standard can be referenced and the "correct" answers can be obtained. Cases can be divided up into the left and right column depending on the gold standard. We can count the number of cases that fall into each of the four resulting categories. A good predictive algorithm should have a high percentage of true positives and true negatives, and few false positives and false negatives.

Another set of measures used commonly in the text mining literature is precision and recall. These measures are very valuable when the positive cases far exceed the negative cases. Under these circumstances a reasonably high specificity can be deceptive. Recall is identical to sensitivity. Precision is the fraction of positive calls that are truly positive:

$$\text{Precision} = \text{TP}/(\text{FP} + \text{TP})$$

These key measures are often used to evaluate different predictive algorithms, and will be used frequently in this book.

2.3 Deriving and analyzing sequences

High throughput sequencing technology has made sequences for genes and proteins readily available (see Figure 2.11), and has permitted the sequencing of the genomes of hundreds of different organisms. The rate at which new nucleotide and amino acid sequences are being produced far exceeds our ability to thoroughly experimentally investigate the genes and proteins they derive from. While direct experimentation is the only absolute way to

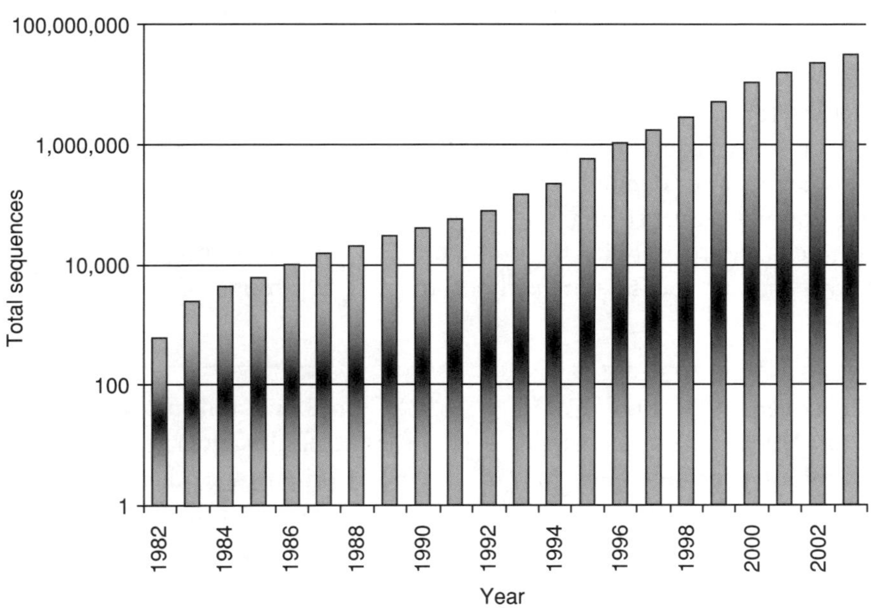

Figure 2.11 *Growth of the GenBank Sequence Database*. The Genbank database has grown dramatically over the last 20 years as sequencing technology has rapidly improved. The *y*-axis is displayed in a logarithmic scale; the growth of the database appears exponential.

understand the function of a gene, analysis of its sequence often provides very good suggestive functions quickly. A host of different sequence analysis strategies have become available, including strategies to compare sequences to one another, to identify common features in families of similar sequences, to identify small patterns in sequences that might indicate functionality, and to cluster similar sequences (Durbin, Eddy et al. 2003). All of these approaches offer the opportunity to transfer knowledge from well-studied genes to newly discovered genes once their sequences become available.

Sequence analysis is relevant to automated mining of the scientific literature in two different ways. The first is using the scientific literature to directly enhance sequence analysis methods. Text mining methods can provide clues about what the common function of the aligned domains is, and can actually be used to improve alignments as well. Second, sequence analysis can be useful in extending text about genes to other genes. Very little is known about most genes; in most cases we are inferring knowledge from a few well-studied genes. Since literature is sparsely available, effective design of functional genomics algorithms requires strategies to make the available literature apply to poorly studied genes. Sequence similarity is one avenue to help us decide which genes correspond with what other genes. If a well-studied gene is similar to a poorly studied one, there is a good chance that they share biological properties and some of the literature from the well-studied gene may apply to the poorly studied one as well.

The content of this section is outlined in the frame box. The first part of this section introduces the reader to the basics of protein and nucleotide sequencing technology. The remainder of the section focuses on sequence analysis techniques. The reader is introduced to pairwise sequence comparison methods first. We describe the standard methods to score sequence alignments, and dynamic programming methods that can be used to find optimal sequence alignments by maximizing that alignment score. We then describe the popular BLAST database query algorithm. The later part of the chapter focuses on methods to analyze groups of sequences. We

1) Sequencing technology
   a) Nucleotide sequencing
   b) Polypeptide sequencing
2) Comparing two sequences
   a) Scoring alignments
   b) Dynamic programming to optimize alignments
   c) BLAST algorithm to query sequence databases
3) Comparing multiple sequences
   a) Multiple sequence alignments
   b) Consensus sequences
   c) Weight matrices
   d) PSI-BLAST
   e) Hidden Markov models

present a discussion of multiple sequence alignment methods, weight matrices to recognizes sequence profiles, PSI-BLAST, and hidden Markov models.

2.3.1 Sequencing

High throughput sequencing technologies have produced a wealth of data in nucleotide sequences and amino acid sequences. High throughput DNA sequencing has fueled the rapid growth of available sequences in GenBank and has been the driving force for efforts such as the human genome project.

DNA is sequenced with the Sanger Dideoxy method depicted in Plate 2.5. It is a fast and simple method that works on a single strand of DNA, and provides the sequence of the reverse complement (Stryer 1995). The sequencing approach uses DNA polymerase to synthesize a complementary sequence. For DNA synthesis to be possible, DNA polymerase must be incubated with the DNA strand and with the four deoxyribonucletide bases that serve as the building blocks for the complementary strand. In addition, fluorescently tagged dideoxyribonucleotide bases are added as well; each base is tagged with a different color. Since they lack the hydroxyl group at the 3′ carbon, the tagged dideoxyribonucleotides are chain terminating nucleotides; they can be appended to the 5′ end of a DNA strand, but cannot be further extended from their 3′. Once DNA polymerase happens to use a dideoxy base instead of the normal deoxyribonucleotide base by chance, the DNA chain becomes terminated. Since each type of didieoxy nucleotide is tagged with a different colored fluorescent dye, the newly synthesized DNA fragment will be tagged with a fluorescent dye the color of which corresponds to the final base in the fragment. Different length fragments are produced, each labeled by a specific color corresponding to the last nucleotide incorporated into the sequence. At this stage the fragments are sorted by length using a technique such as electrophoresis or chromatography and run through a detector. The sequence of colors corresponds to the DNA sequence.

Shotgun assembly strategy is the preferred approach for modern genome sequencing. Genomic DNA is decomposed into random small overlapping fragments; the Sanger dideoxy sequencing technique is used to rapidly sequence these fragments. Genome assembly algorithms piece together the resulting small sequences to obtain the sequence of the original genome.

Protein sequences are determined using Edman degradation (Figure 2.12). Sequencing proceeds from the N-terminal side of the protein. Phenyl isothiocyanate is a compound that reacts specifically with the amino group of the N-terminal amino acid of the protein. Acidification of the protein solution causes removal of the N-terminal amino acid bound to phenyl isothiocyanate. The single amino acid bound to phenyl isothiocyanate is then identified through high pressure liquid chromatography (HPLC).

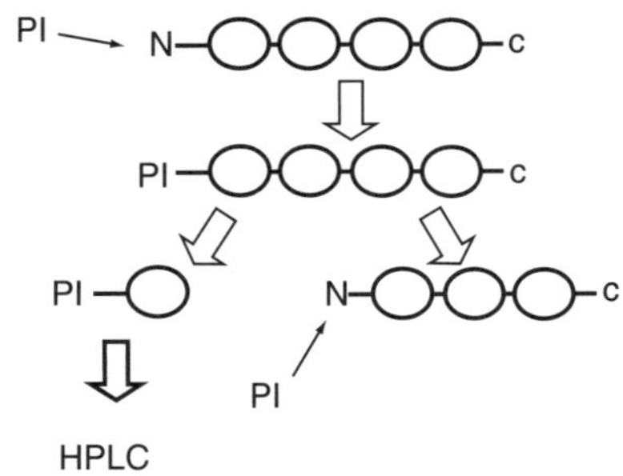

Figure 2.12 *Edman reaction for protein sequencing.* Phenyl isothiocyanate (PI) is used in a sequence of reactions known as Edman degredation to sequence proteins. Phenyl isothiocyanate reacts and binds the N-terminal end of the protein. Acidification then causes the detachment of the terminal amino acid, which can then be identified by high pressure liquid chromatography (HPLC).

Removal of the N-terminal amino acid exposes a new N-terminal amino acid. The process is then repeated on the remaining protein, to identify the next amino acid.

Unlike DNA sequencing, protein sequencing is a relatively costly and laborious process. In practice most protein sequences are inferred from DNA sequences translated through the genetic code.

2.3.2 Homology

While significant amounts of protein and nucleotide sequence information is currently available, little is known about most genes. While it is desirable to directly study all genes experimentally, detailed experiments for each gene in each organism is not feasible. Fortunately, genes and proteins are often related to one another through evolution. We refer to such pairs of genes as homologous. Genes with similar sequences are often evolutionarily related, and have similar functions that have been conserved; therefore we can guess the function of a gene by finding other similar genes that have been well studied. Computational algorithms can help us assess whether sequences are sufficiently similar to be homologous. For a more comprehensive discussion of this subject the reader is referred elsewhere (Durbin, Eddy et al. 2003).

In the remainder of this section we will discuss protein sequence comparison and analysis; however, everything that is presented can be applied to nucleotide sequences as well without any difficulty.

We seek to address the question of similarity between two sequences. The most straightforward strategy to look for similarity between sequences is to look for exact matches between subsequences. Under these circumstances the challenge of homology searching is reduced to string matching. The first difficulty with this strategy is that matches are rarely exact, and there are frequently minor substitutions, insertions and deletions that have occurred in biology. For instance, two sequences might be related evolutionarily and may be very similar, but chance may have deleted a five amino acid loop in the second sequence without changing the structure or function of the protein significantly. Additionally, since certain amino acids are relatively similar to each other, certain particular amino acids can substitute for others without significantly affecting the structure or function of the protein either. These substitutions are referred to as conserved substitutions. Since matches are rarely exact, a more flexible approach must be taken.

Historically, biologists used dot plots to manually identify regions of similarity in two protein sequences despite insertions, deletions, and substitutions (see Figure 2.13). In a dot plot the two sequences that are being assessed for homology are arranged in a matrix so that the first sequence is running down the left side, while the second is running across the top. The row of a box in the matrix corresponds to a position in the first sequence; the column that a box is in corresponds to a position in the second sequence. A dot is placed in each box where the amino acids or nucleotide

	A	A	C	T	T	T	G	G	A	G	A	A	T
A	♦	♦							♦		♦	♦	
A	♦	♦							♦		♦	♦	
C			♦										
T				♦	♦	♦							♦
T				♦	♦	♦							♦
G							♦	♦		♦			
C			♦										
G							♦	♦		♦			
A	♦	♦							♦		♦	♦	
G							♦	♦		♦			
A	♦	♦							♦		♦	♦	
A	♦	♦							♦		♦	♦	
C			♦										

Figure 2.13 *Using the dot plot to compare sequences.* Two similar sequences can be compared with the dot plot. One sequence is listed across the top, with the second along the side of the matrix. Dots are placed in the boxes where the corresponding nucleotides or amino acids are identical. Diagonals indicate identical regions in the sequence.

bases from both sequences in corresponding positions are identical. The pattern of dots allows regions of similarity to become visually apparent.

2.3.3 Sequence alignment

Sequence alignment is the computational approach to comparing sequences. Alignment algorithms strive to arrange the sequences alongside each other in a way so that they match up with each other as best as possible. The goal of pairwise alignment is to compare two sequences to each other according to their similarities, while multiple alignment strives to organize a family of sequences so that their commonalities become apparent. In Figure 2.14 we present an example of two aligned sequences. The bars (|) between the sequences indicate identical matches, while the colon (:) indicates functionally conserved amino acids. The dashes in the sequences indicate that a gap was inserted in that sequence to allow it to match up appropriately with the other sequence.

Critical to any algorithm that aligns sequences is developing a scoring function that is proportional to the quality of the alignment that the algorithm can optimize. Such a scoring function must reward matches between the sequences, and penalize mismatches. Since insertions and deletions are possible evolutionarily, alignments and scoring functions must account for the possibility of gaps. In general, these scoring systems are additive functions; there are positive terms for matches, and negative terms for mismatches and gaps.

Some amino acid substitutions are more likely than others. In fact amino acids that are physiologically similar may replace each other during evolution with relative frequency compared to dissimilar amino acids. A good scoring function should not penalize all mismatches equally, but rather should take into account the probability that a particular substitution has occurred in the context of evolution. Investigators have devised substitution matrices that assign a score to each pair of amino acids; the scores for each pair are based on the ratio of the probability that two amino acids could

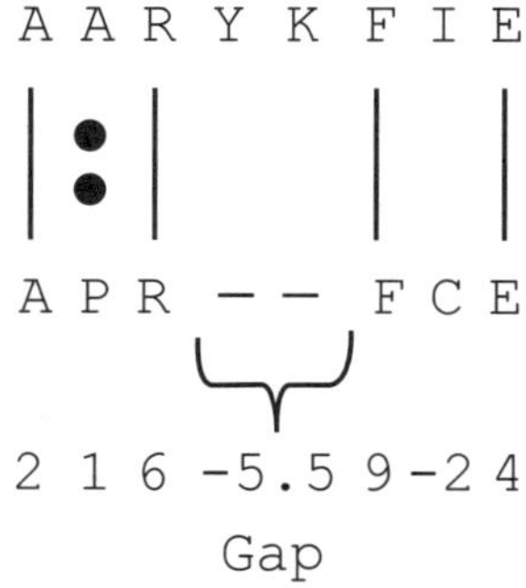

Figure 2.14 *Example of two short aligned sequences*. The bars between the sequences indicate identity at that position. The dots between the two sequences indicate similar but not identical amino acids. The substitution score between the amino acids at each position is listed. Also a gap is placed in the second sequence to optimize the alignment. The penalty score of the gap in this example is −5.5. The total score of an alignment is the sum of the substitution scores and the gap penalty. In this case the score is 14.5.

have replaced each other during evolution to the probability that the same two amino acids are observed at random in two unrelated sequences. A larger score between two amino acids indicates the possibility that the substitution is a functionally conserved position.

Assuming that the probability of each of the amino acids is independent, given two aligned sequences x and y where the amino acid in the i-th position of the alignment for sequence x is x_i, the probability that an alignment is a random occurrence can be represented as follows:

$$Q_{x,y} = \prod_i q_{x_i} q_{y_i}$$

where q_{x_i} is the frequency of amino acid i. We are assuming that the amino acids at each position are entirely independent. Similarly the probability that the alignment is a consequence of evolution can be represented as

$$P_{x,y} = \prod_i p_{x_i, y_i}$$

where p_{x_i, y_i} is the probability that amino acids x_i and y_i are derived from an unknown, but common parent. The log likelihood ratio of these two models is:

$$\log\left(\frac{P_{x,y}}{Q_{x,y}}\right) = \sum_i \log\left(\frac{p_{x_i, y_i}}{q_{x_i} q_{y_i}}\right)$$

This is an alignment score. Substitution matrices contain estimates of the log term for each of the possible amino acid pairs. For each pair the approximate frequencies of amino acids are used to estimate q and the estimated probability of a substitution between two specific amino acids in evolutionarily related sequences is used for p. One substitution matrix, the PAM250 matrix, is depicted in Figure 2.15 (Pearson 1990). Substitution matrices are used to score matches and mismatches during sequence alignment. The alignment score is the sum of these matrix terms.

Gaps in alignment are used to account for possible insertions and deletions during evolution. Typically gaps are penalized as a function of only their length, and not of their content. A common form for gap penalties is the affine gap penalty function:

$$\gamma(g) = \delta - (g - 1)\varepsilon$$

where g is the length of the gap, δ is a penalty for opening a gap, and ε is a linear penalty for extending the gap. The gap-opening penalty insures that alignments are penalized for having an excessive number of individual gaps.

	A	R	N	D	C	Q	E	G	H	I	L	K	M	F	P	S	T	W	Y	V
A	2																			
R	–2	6																		
N	0	0	2																	
D	0	–1	2	4																
C	–2	–4	–4	–5	4															
Q	0	1	1	2	–5	4														
E	0	–1	1	3	–5	2	4													
G	1	–3	0	1	–3	–1	0	5												
H	–1	2	2	1	–3	3	1	–2	6											
I	–1	–2	–2	–2	–2	–2	–2	–3	–2	5										
L	–2	–3	–3	–4	–6	–2	–3	–4	–2	2	6									
K	–1	3	1	0	–5	1	0	–2	0	–2	–3	5								
M	–1	0	–2	–3	–5	–1	–2	–3	–2	2	4	0	6							
F	–4	–4	–4	–6	–4	–5	–5	–5	–2	1	2	–5	0	9						
P	1	0	–1	–1	–3	0	–1	–1	0	–2	–3	–1	–2	–5	6					
S	1	0	1	0	0	–1	0	1	–1	–1	–3	0	–2	–3	1	3				
T	1	–1	0	0	–2	–1	0	0	–1	0	–2	0	–1	–2	0	1	3			
W	–6	2	–4	–7	–8	–5	–7	–7	–3	–5	–2	–3	–4	0	–6	–2	–5	17		
Y	–3	–4	–2	–4	0	–4	–4	–5	0	–1	–1	–4	–2	7	–5	–3	–3	0	10	
V	0	–2	–2	–2	–2	–2	–2	–1	–2	4	2	–2	2	–1	–1	–6	0	–6	–2	4

Figure 2.15 *Example of a substitution matrix*. The PAM250 substitution matrix is displayed above. The amino acids are listed by their single letter designation across the top and the sides of the matrix. Scores are higher for functionally similar amino acids.

That is, the algorithm is biased towards continuous gaps, and away from multiple short discrete gaps.

In Figure 2.14 we demonstrate how we can use these concepts to score an alignment based on the addition of substitution scores and gap penalties.

2.3.4 Pairwise sequence alignment and dynamic programming

The dynamic programming approach can be used to find the optimal alignment that maximizes the alignment score presented in the last section. Dynamic programming is a general computer science term that applies to solving and storing the solutions to smaller sub-problems in order to solve larger problems. Needleman and Wunsh introduced this concept to sequence alignment (1970). The idea of dynamic programming here is to use alignments of smaller subsequences to determine the alignment of the larger sequence (see Figure 2.16).

Say there are two sequences x and y, where x_i represents the amino acid in sequence x at position i, and y_j represents the amino acid in sequence y at position j. Now, consider a subsequence of x consisting only of amino acids $1, \ldots, i$, and a subsequence of y consisting only of amino acids $1, \ldots, j$. We use the term $S_{i,j}$ to indicate the score of best possible alignment of these two subsequences in which x_i and y_j are aligned next to each other. Such an alignment may have gaps prior to x_i and y_j positions, but may not end on a gap. We create a matrix, S, where the $S_{i,j}$ scores will be calculated for all positions i and j and then maintained (Figure 2.17).

```
A           AARYKF
A           APR--F

AA          AARYKFI
AP          APR--FC

AAR         AARYKFIE
APR         APR--FCE
```

Figure 2.16 *Aligning subsequences.* Here we list the smaller alignments of the two aligned sequences in Figure 2.14. Alignments between these smaller sequences can be built up into larger alignments in dynamic programming.

	A	A	R	Y	K	F	I	E
A	2	2	–2	–3				
P	1	3	2	–7				
R	–2	–1	9	–2				
F	–4	–6	–5	16				
C					11			
E								

Figure 2.17 *Dynamic programing score matrix.* The above matrix shows scores for optimal alignments for subsequences. For example, the score of the optimal alignment between "AARY" and "AP" ending with P and Y aligned is listed in the forth column and the second row; the score of that alignment is -7. As an illustrative example in calculating alignment scores, consider the optimal alignment between the sequences "AARYK" and "APRFC" where "K" and "C" are aligned with each other. The score of that alignment is equal to the sum of the substitution score between K and C (-5), one of the scores in the blackened boxes, and the appropriate gap penalty. Those scores in the blackened boxes are optimal alignments of other possible subalignments. If there is no gap whatsoever the corner box is used with a score of 16 and no gap penalty is added. If there is a gap introduced in the sequence along the top, a score from the blackened column is taken and the gap penalty is added. If a gap is introduced in the sequence along the left, a score from the blackened row is taken and a gap penalty is added. The box from which the highest score is derived is noted and stored. That highest possible score is entered as the optimal score.

If the best scoring subsequence alignment has no gaps in the positions right before x_i or y_j, then the score of this alignment should be the score of the optimal alignment up to the positions right before i and j, plus the reward or penalty of having amino acid x_i and y_j aligned against each other. That is:

$$S_{i,j} = S_{i-1,j-1} + s(x_i, y_j)$$

where $s(x_i, y_j)$ is the substitution matrix score of amino acids x_i and y_j.

Now, instead say that in the best alignment there is a gap of length g right before x_i. Then the last position before i and j where two amino acids are aligned is $i-1$ in x and $j-1-g$ in y. This is because a gap of length g skips over that many positions in sequence y. The subsequences up to these two amino acids preceding the gap must be aligned in the best possible way. The score of the best possible alignment between the subsequence of x up to $i-1$ and the subsequence of y up to $j-1-g$ is $S_{i-1,\,j-1-g}$. Under these circumstances the score of the best possible alignment up to i and j must be $S_{i-1,\,j-1-g}$, plus the substitution score between x_i and y_j plus the gap penalty:

$$S_{i,\,j} = S_{i-1,\,j-1-g} + \gamma(g) + s(x_i,\, y_j)$$

where $\gamma(g)$ is a gap penalty of length g. Alternatively, if there is a gap of length g right before y_j, then similarly:

$$S_{i,\,j} = S_{i-1-g,\,j-1} + \gamma(g) + s(x_i,\, y_j)$$

Now, if we are trying to calculate $S_{i,\,j}$ and we do not know if there is a gap in x or y preceding the i and j position, then we can consider all of the possible scenarios: (1) no gap, (2) gaps of any length possible in x, and (3) gaps of any length possible in y. The correct situation is the one that derives the largest value of $S_{i,\,j}$—that is the best possible alignment of the two substrings. So, since $S_{i,\,j}$ is the highest scoring alignment, then it has to be the case that:

$$S_{i,\,j} = \max\begin{cases} S_{i,\,j} = S_{i-1,\,j-1} + s(x_i,\, y_j) & \\ S_{i,\,j} = S_{i-1,\,j-1-g} + \gamma(g) + s(x_i,\, y_j) & 1 \le g < j\text{ -1} \\ S_{i,\,j} = S_{i-1-g,\,j-1} + \gamma(g) + s(x_i,\, y_j) & 1 \le g < i\text{ -1} \end{cases}$$

Therefore, $S_{i,\,j}$ can be calculated if all $S_{1\ldots i-1,\,j-1}$ and $S_{i-1,\,1\ldots j-1}$ are known (see Figure 2.18). Each score $S_{i,\,j}$ is derived from one of these prior values in the matrix. The prior position tells us if there is a gap, how long the gap is, and what the previous alignment is. So we maintain a pointer to the cell in S from which $S_{i,\,j}$ was calculated.

Dynamic programming works by calculating each of the scores $S_{i,\,j}$ for the matrix. The values in the first row and column are set to the substitution matrix values for the amino acids they correspond to; these are just single amino acid alignments. Then the other matrix values are calculated iteratively using the procedure described above (see Figure 2.17). This process is repeated until all scores S are determined.

The score of the best alignment between sequences x and y is the highest score in the last column or last row of matrix S. To determine the best

	A	A	R	Y	K	F	I	E
A	2	2	–2	–3	–1	–4	1	0
P	1	3	2	–7	–4	–6	–6	–2
R	–2	–1	9	–2	0.5	–7	–5.5	–5
F	–4	–6	–5	16	–1	12.5	4	–2.5
C	–2	–6	–6.5	4	11	7	10.5	5
E	0	–2	–4	–/5	11	6	5.5	14.5

Figure 2.18 *Tracing back during alignment with dynamic programming.* To obtain the final sequence alignment score, first the score for the entire alignment matrix is calculated. Then the highest score in the final row or column is identified. In this case it is 14.5. Since we stored the boxes from which each alignment score derived, we can trace back to the beginning of the sequences. As we go, we can write out the alignment. Each box in the path indicates which amino acids in the sequence are aligned in the final alignment. For example, in this case the score of the alignment comes from the box in the last row and column. That box also tells us that the final position of the alignment involves E and E being aligned to each other. Then we trace back to the prior box. There is no gap, and we see that going backwards I and C are aligned in the final alignment. When the trace back goes up and across a single box, no gap is introduced. When the trace back skips across boxes, a gap is introduced.

alignment between x and y, go to that cell in matrix S and use the pointers to trace back the route that was taken to get there (see Figure 2.18). The coordinates of each cell on the path indicate the amino acids in x and y that align together. If, for instance, the path does through $S_{m,\,n}$ then amino acids x_m and y_n are aligned in the optimal alignment; where the pointer proceeds in a diagonal there is no gap.

This approach gives an exact solution for the optimal alignment. The dynamic programming approach is effective because of the additive nature of the scoring function. The contributions of the scoring function can be easily decomposed into individual parts.

2.3.5 Linear time pairwise alignment: BLAST

While dynamic programming provides the optimal alignment for any two given sequences, it is poorly suited for searching databases of sequences. Dynamic programming with affine gap penalties is extremely sensitive and is generally regarded as the gold standard for sequence similarity. But the difficulty is that it requires polynomial time. In its fastest implementation, given two sequences of length n it requires a number of computations proportional to n^2 (Gotoh 1982). For large database searches against millions of sequences, this method is often too slow.

Linear time algorithms that rely on heuristics have been introduced. These methods are considerably faster, and are well suited for querying sequences against large databases (Pearson and Lipman 1988; Altschul, Gish et al. 1990; Pearson 1990). They give approximate solutions, but often the results are quite compelling, and they frequently select the most similar sequences. The most popular of these methods is the Basic Linear Alignment Search Tool, or BLAST; it is available for both nucleotide and amino acid sequences (Altschul, Gish et al. 1990).

The principle behind BLAST is that if two sequences are homologous, they will likely have short stretches, or words, that are either exact matches or at least high scoring matches. Given a query sequence, BLAST creates a list of "neighborhood words", which are short sequences of a fixed length (3 for amino acids, 11 for nucleotides) that have a similarity score exceeding a certain pre-selected threshold to the query sequence.

The algorithm then searches for sequences that have an exact match to the neighborhood words in the database. Such a matching sequence is referred to as a hit. Once a hit is obtained, BLAST attempts hit extension; that is, it extends the alignment without gaps until the maximum possible score is achieved. It stops extension when the score of the extension falls below the maximum score observed during extension by a certain threshold.

BLAST has become one of the most common bioinformatics tools used by biologists worldwide. It is a standard tool used by investigators trying to assess the function of a newly discovered gene or protein.

2.3.6 Multiple sequence alignment

While pairwise sequence comparison is an extremely powerful tool, there are many situations in biology where a sequence has to be considered in the context of a whole family of proteins. In general, genes and proteins come as classes. Often, a gene will have many known cousins that are related through evolution or share common functionality. Under these circumstances it is critical to be able to appreciate the common conserved features between these sequences as they give insight as to which parts of a protein are critical for structure and function. For example, the binding site that is conserved in a family of related proteins across many species is likely to be conserved in a similiar query sequence as well. Moreover, the common features may indicate the defining aspects of that family of proteins; this may allow us to predict whether newly discovered sequences belong to the family as well.

An example of multiple alignment is presented in Figure 2.19, made by the algorithm Clustal W (Thompson, Higgins et al. 1994).

As with pairwise alignment, it is critical to have a scoring method for multiple alignment. While making multiple alignments, one would prefer

```
b  MSVMYKKILYPTDFSETAEIALKHVKAFK-TLKAEEVILLHVIDEREIK----------- 48
c  --VMYKKILYPTDFSETAEIALKHVKAFK-TLKAEEVILLHVIDEREIKVEEFENELKNK 57
a  MPVAKDNQFW-DALMEN-KVAKKLIKKHKCKAKCENID--DLANRYEVS----------- 45
     *  .: ::   : *. ::* * :* .* . *.*::   .: :. *:.
```

Figure 2.19 *Multiple sequence alignment.* A multiple alignment between three sequences. This alignment was generated by Clustal W.

columns in the alignment where all of the sequences have the same or functionally conserved amino acids. One commonly employed strategy is to use a function that is proportional to the uniformity of a column, and then sum up the scores for each of the columns. A common scoring system for a column is the sum of pairs scoring system. Given a column of amino acids in a multiple alignment m, its score is the sum of all of the pairwise substitution scores in the column:

$$S(m) = \sum_{i<j} s(m_i, m_j)$$

where i and j are indices for sequences. Here the score s is obtained by referencing a substituion matrix, like the PAM matrix. The score of an entire alignment can be calculated by summing all the scores for each of the independent columns.

Similar to pairwise alignment, a gap penalty needs to be included to weight alignments away from those that create excessive numbers of gaps. We can use the exact same sort of penalty function used in pairwise alignment.

Dynamic programming can be used to maximize scoring schemes like the one introduced above. Dynamic programming is used to find optimal multiple alignments using these scoring schemes in a similar fashion as it is used to find pairwise alignments. The difference is that instead of a two-dimensional scoring matrix, we need to implement an n-dimensional matrix where each dimension represents each of the different sequences. However, the amount of memory and time required for this approach is prohibitive for more than a small number of sequences. The memory required is proportional to n^m if each sequence is of length n and there are m sequences. The number of computations necessary is proportional to $2^m n^m$.

Progressive alignment methods are most commonly used to obtain multiple alignments. These, and other commonly used alignment methods, are heuristic methods. These methods work well in practice and they are often efficient. Progressive alignments work by first using dynamic programming to do all pairwise alignments. Then the closest two sequences are used as a starting point, and additional sequences are added to the alignment iteratively until all of the sequences have been aligned.

2.3.7 Comparing sequences to profiles: weight matrices

Once a multiple alignment is created, it is possible to create a profile that represents the common features across the sequences. Such a profile can be used to scan new sequences and to ascertain whether they may be functionally or evolutionarily related to the sequences that the profile represents.

The traditional approach to this problem is the consensus sequence. Consensus sequences are short sequences, where for each position the most common amino acids (or nucleotides) that occur are listed. In Figure 2.20 we list several observed nucleotide sequences and a corresponding consensus sequence. This is a commonly observed eukaryotic promotor sequence known as the CCAAT promoter (Mantovani 1998). Consensus sequences have been effective in defining promoter regions in gene sequences and functional motifs in protein sequences. One of the shortcomings of consensus sequences is that it accounts poorly for frequency information. For example, a nucleotide sequence may have G in the first position 80% of the time, and T 20% of the time. The consensus sequence for the first position could be written as G/T; in this case the presence of either nucleotides would be regarded as equally acceptable. This is clearly not the case as G is more frequently observed. However, if we write only G in the consensus sequence then we ignore all the cases with T in the first position.

Observed sequences

1	2	3	4	5	6	7	8
C	C	A	A	T	C	C	C
C	C	A	A	T	C	A	C
C	C	A	A	T	C	A	C
C	C	A	A	T	C	G	T
C	C	A	A	T	C	G	T

Consensus sequences

C	C	A	A	T	C	A/G	T/C

Figure 2.20 *Using consensus sequences to summarize multiple alignments.* A collection of five aligned sequences each representing a CCAAT promotor is depicted. The consensus sequence is created by noting the most frequent amino acids at each position.

Another more sophisticated strategy is the weight matrix. It is similar to consensus sequences in that it has information about short sequences in a position specific manner. However, weight matrices maintain frequency information for each position. Let us suppose we have a collection of N sequences of length n that are known to be in the same class, C. The weight matrix, W, is calculated from the collection of N sequences. The weight matrix, W, has n columns, one for each sequence position, and 20 rows, one for each amino acid. Given a new sequence of length n, we can apply the weight matrix to obtain a score proportional to the chance that the sequence is in class C (Figure 2.21). For each position in the new sequence identify the amino acid. Then find the value of the weight matrix corresponding to that amino acid at the column representing the position. Then sum these weights to score the sequence.

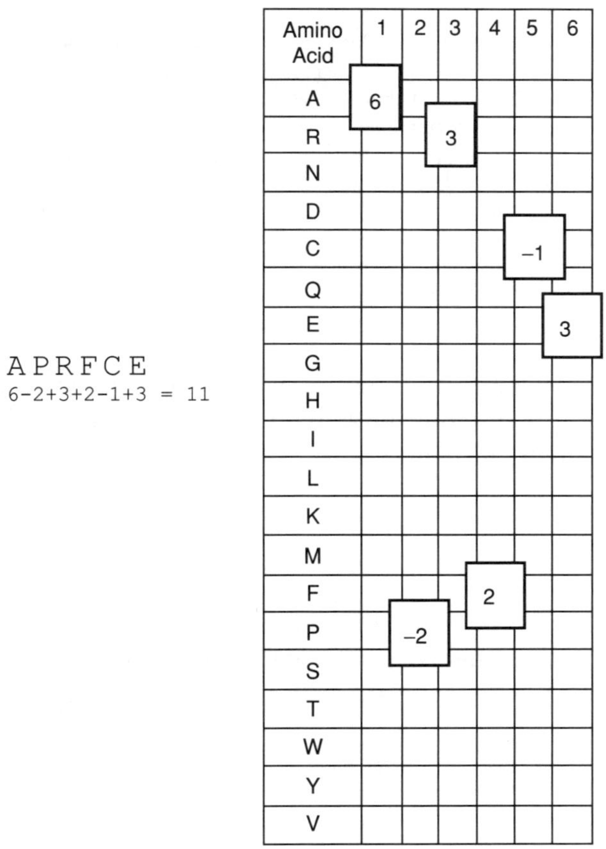

Figure 2.21 *Using a scoring matrix to score a sequence against a multiple alignment.* A multiple alignment is used to construct a weight matrix. Each box in the matrix corresponds to a position in the alignment (column) and an amino acid (row). It contains a value that is proportional to the likelihood that a sequence from the multiple alignment has that amino acid at that position. So a sequence of amino acids can be scored by adding up the corresponding matrix values at each position. The larger the score, the more likely it is similar to the multiple alignment sequences.

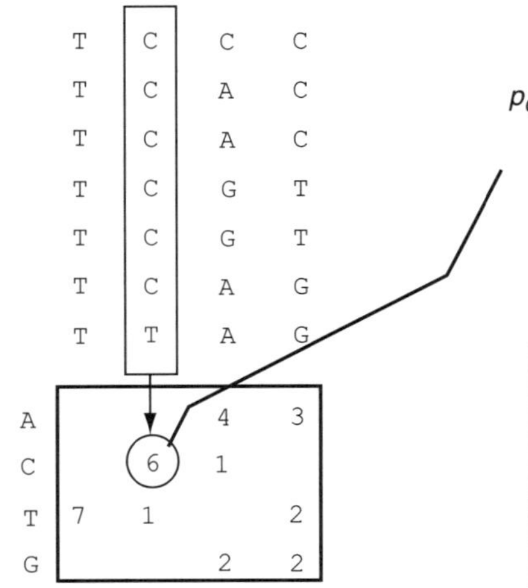

$$p_{C,2} = \frac{N_{C,2} + q_C}{N + 1} = \frac{6 + 0.25}{7 + 1} = 0.79$$

$$\log_2 \left(\frac{p_{C,2}}{q_C} \right) = 1.64$$

	1	2	3	4
A	−3	−3	1.1	0.7
C	−3	1.6	−0.7	−3
T	1.9	−0.7	−3	0.2
G	−3	−3	0.2	0.2

Figure 2.22 *Creating a weight matrix.* This is an example of how to construct a weight matrix. Given a collection of sequences we total up the number of each nucleotide or amino acid observed in each position. To calculate the probability of each nucleotide or amino acid at each position we divide the number of times a particular nucleotide is observed at a position by the total number of sequences. Here we consider the probability of the second position having nucleotide C, $p_{C,2}$. Often, it is advantageous to include pseudo-counts. We assume the extra counts are proportional to the background frequency of nucleotides, q_C; in this case we assume that each nucleotide has a background frequency of 0.25. A pseudo-count of one is used here. This avoids zero probabilities. We divide the probability of each nucleotide at each position by the background probability of the nucleotide. The log of that ratio is the weight matrix value.

Figure 2.22 demonstrates how the weights are obtained for a weight matrix. For each position the probability that a particular amino acid is at that position can be calculated:

$$P_{a,i} = \frac{N_{a,i}}{N}$$

where $P_{a,i}$ is the probability of amino acid a in position i, N is the total number of sequences, and $N_{a,i}$ is the number of sequences with amino acid a in position i. Often pseudo-counts must be used to avoid extreme probabilities. In this case

$$P_{a,i} = \frac{N_{a,i} + q_a}{N + 1}$$

where q_a is the background frequency of amino acid a. This assumes we have a single extra sequence that has amino acids distributed according to the normal background frequency of amino acids.

The values of the weight matrix can be calculated as follows

$$W_{a,i} = \log\left(\frac{P_{a,\,i}}{q_a}\right)$$

Given an unknown sequence, s, of length n we can use it to calculate the log likelihood that it belongs to the same class using the weight matrix. If we assume that each of the positions are independent of one another:

$$\log\left(\frac{P(s|C)}{P(s|\sim C)}\right) = \log\left(\frac{\prod_i p(s_i,\, i|C)}{\prod_i q(s_i)}\right)$$

where $P(s|C)$ is the probability of the sequence assuming it is in class C, and $P(s|\sim C)$ is the probability of the sequence assuming it is not in class C, then $P(s_i, i|C)$ is the probability that the amino acid s_i would be at position i in sequence s, if s was in class C. We can decompose this equation into

$$\sum_{i=1}^{n} \log\frac{p(s_i,\, i|C)}{q(s_i)}$$

This is the same as:

$$\sum_{i=1}^{n} W_{s_i,\, i}$$

The log likelihood that any sequence belongs to the class C is simply the sum of the weight matrix terms from each column that corresponds to the amino acid in that position in the sequence.

2.3.8 Position specific iterative BLAST

Position specific iterative BLAST or (PSI-BLAST) is a more sensitive version of BLAST that combines the speed of BLAST with the sensitivity of weight matrices (Altschul, Madden et al. 1997). It is depicted in Figure 2.23. It uses principles of multiple alignment in its database search.

Given a sequence, PSI-BLAST first runs a regular BLAST search to find a collection of similar sequences. These sequences are collected together into an ungapped multiple alignment with the query sequence. The multiple alignment is then used to construct a profile that is similar to the weight matrix described in the previous section. This weight matrix has the same number of columns as the query sequence. PSI-BLAST then uses the profile to run BLAST again and query the database a second time. To do this with a

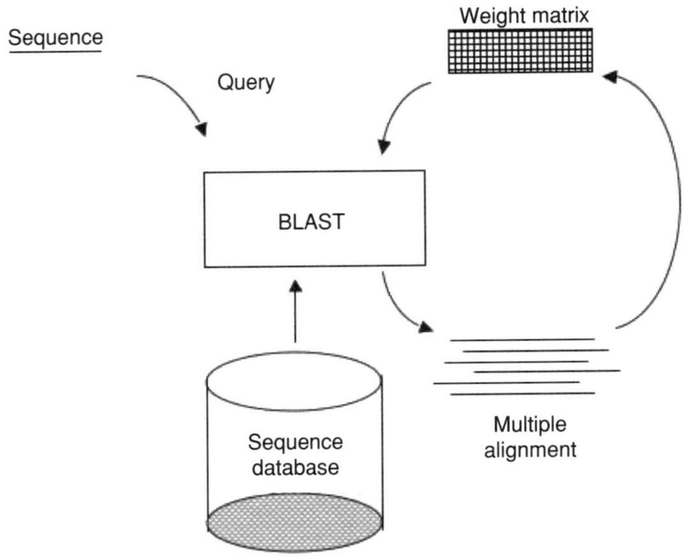

Figure 2.23 *Schematic of PSI-BLAST.* Given a sequence, PSI-BLAST operates by first running a BLAST search. The resulting sequences are organized into a multiple alignment. This multiple alignment is used to generate a weight matrix. The weight matrix is used in another BLAST search, and the weight matrix is updated with the obtained sequences.

profile, a list of neighborhood words of a fixed length that match the profile must be constructed. This is done by identifying words that match any position in the weight matrix at or above a specific threshold. This process is repeated until the algorithm converges or fails to converge after a fixed number of iterations.

The purpose of the iterative approach is that the query sequence is part of a larger family of homologous sequences. By iteratively searching the PSI-BLAST database, the hope is that we are building a profile that corresponds to the family that the query sequence is a part of, and retrieving all of the sequences in that family in the process.

One of the difficulties with PSI-BLAST is sequence contamination. That is if erroneous sequences are retrieved during database search, they could corrupt the profile, and draw in other erroneous sequences. This could potentially have the effect of causing the algorithm to converge on the wrong sequence family or fail to converge altogether.

In general PSI-BLAST is regarded as being more sensitive then BLAST, though it is somewhat slower.

2.3.9 Hidden Markov models

An extremely valuable sequence analysis method is the hidden Markov model (Krogh, Brown et al. 1994). In this section we will be talking about it in the context of multiple sequence alignment, but it has wide application to many areas of sequence analysis as well as text analysis. The hidden Markov model (HMM) is a probabilistic model that assumes that a sequence of observations are accounted for by hidden states. In this case our sequence of observations are nucleotides in DNA or RNA or amino acids observed in a protein. The idea is that the probability of a particular observation depends only on the state of the system. An additional assumption is the Markov assumption; that is the probability of the state of the system is dependent only on the prior state of the system, and does not have any dependence on any other previous state.

As a simple example to illustrate a hidden Markov model, consider a sequence of coin flips. For each flip we observe tails (T) or heads (H). Now let's assume that the flips can be derived by either a fair coin, in which the probability of both H or T is 0.5, or a biased coin in which the probability of H is p and T is $1 - p$. Now assume that between flips the coin can be switched with some probability. The fair coin can switch to the biased one with probability q, while the biased coin can switch to the fair one with probability q'. If we were to model this situation with a HMM, our hidden state would be the coin. There are two possible states, one in which the flips are generated by the fair coin (*S1*) and a second in which flips are generated by the biased coin (*S2*). The probability of H or T at any given point depends on whether *S1* or *S2* is the current state. This simple two-state HMM is illustrated in Figure 2.24. The arrows indicate state transitions,

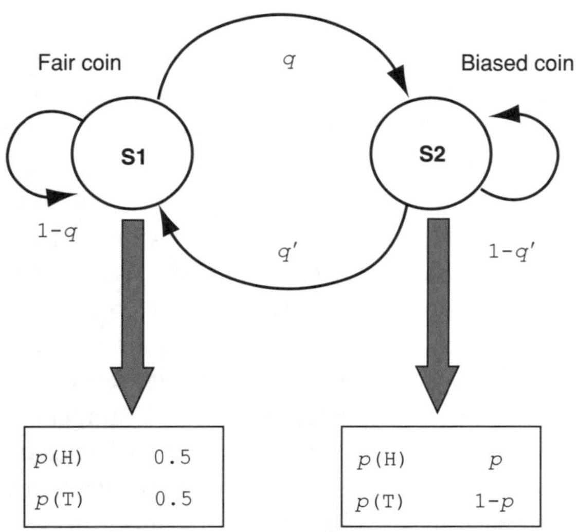

Figure 2.24 *An example of a hidden Markov model.* This is a simple hidden Markov model with two states. The fair coin state, S1, generates heads and tails with equal probability. That state can transition to state S2 after each flip with probability q. The biased coin state, S2, generates heads with probability p.

and the probabilities that those transitions can occur are listed next to the arrow.

Imagine that we have a sequence of observations, and their hidden states are known. Consider the n-th observation. Let us assume that the state is known to be S_n at that point and the observation O_n is observed. The probability of an observation depends only on the state that it is in at the time. So, the probability of the observation, $P(O_n|S_n)$, is a parameter in the hidden Markov model. The probability of the state is dependent only on the prior state. So, the transition probabilities between the hidden states, $P(S_n|S_{n-1})$, are also model parameters. So given a sequence of observations, S, the probability of that sequence of states and observations can be calculated. In general:

$$P(S) = P(O_1 \ldots O_n) = P(O_1)P(O_2|O_1)P(O_3|O_1,O_2)\ldots P(O_n|O_1 \ldots O_{n-1})$$

But the hidden Markov model assumptions allow us to make the simplification that the probability of each observation depends only on the state of the system when it was generated. In addition the probability of each state depends only on the prior hidden state. So the probability of a sequence of observations is the probability of the sequence of hidden states multiplied by the probability of the observations given those hidden states. Therefore the probability of a sequence if the hidden states are known is:

$$P(S) = P(O_1 \ldots O_n, S_1 \ldots S_n) = \prod_{i=1}^{n} P(O_i|S_i)P(S_i|S_{i-1})$$

In the case of sequence analysis, instead of observing heads and tails on a coin, we observe nucleotides and amino acids. In these ċases the state is a physiologic state. For example, hidden Markov models are often used as a generalization of the weight matrices introduced in the previous example. Each state might represent a particular site in a class of proteins. Each site would have a biased distribution for amino acids. Each site might transition to the next site in the sequence that we would expect. However, it might with a finite probability also skip the next site (a deletion), or it might have an extra amino acid in place prior to the next site (an insertion). A typical scheme for a hidden Markov model for sequence alignment is demonstrated in Figure 2.25. Hidden Markov modes can also be used for secondary structure prediction; a typical Markov model is demonstrated in Figure 2.26. There are three states: alpha helix, beta sheet, and loop. All amino acids have one of the three hidden states. Any of the states can transition to any of the others.

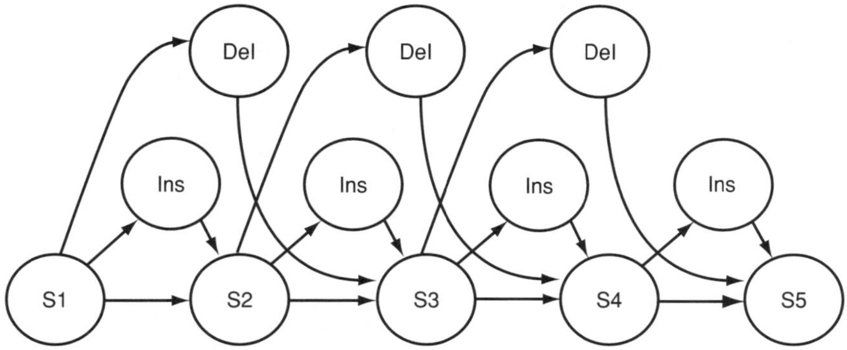

Figure 2.25 *Example of a hidden Markov model to align sequences.* In this schematic we see five positions in the alignment: S1 to S5. Each position generates amino acids with a particular probability distribution. Each position can transition to the next. In addition deletion (Del) states allow one of the positions to be skipped. Insertions (Ins) states allow for amino acids to be generated between the prespecified states.

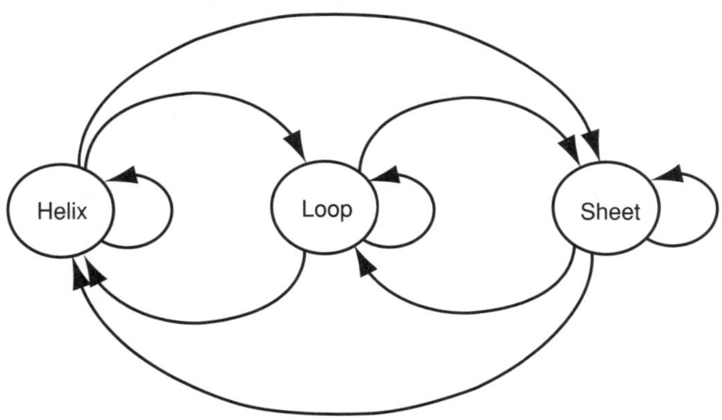

Figure 2.26 *Example of a hidden Markov model to predict secondary structure.* There are three possible secondary structure states: *Helix*, *Loop*, and *Sheet*. Each one generates amino acids with different probabilities. Each state can transition to any of the other two with a certain probability.

Given a hidden Markov model whose parameters have been fully determined we can determine the most likely underlying hidden states for a given sequence of observations. In the case of secondary structure prediction, for example, this would be tantamount to determining which amino acids in a protein are beta sheets, alpha helices, or loops. Similarly in the case of sequence alignment, the underlying state determines whether each amino acid corresponds to a particular position in the alignment or is in an insertion or deletion.

Dynamic programming can be used to calculate the most likely path through the hidden states given a series of observations. This algorithm is known as the Viterbi algorithm and is demonstrated in Figure 2.27.

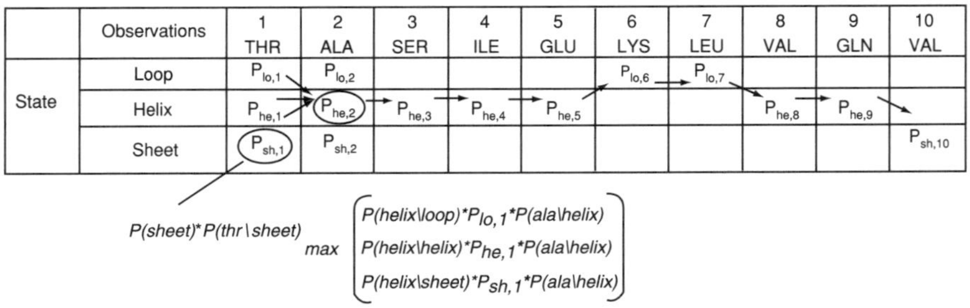

Figure 2.27 *The Viterbi algorithm.* The Viterbi algorithm allows us to assign the most likely states to a sequence of observations. At each position in the sequence we calculate the probability of the highest probability path ending with each state. An example of that calculation is depicted above. For each position and state we note the prior state used to generate the probability at that position. Upon completion, we can trace back through the sequence to identify the highest probability pathway through the hidden states.

In this algorithm we start at the beginning of the sequence and at each position calculate the probability of the most likely state while remembering the paths that we went through to get there. The key is that the optimal path to get to position k is related to the optimal path to get to position $k - 1$.

Say that $p_{i,\,k}$ is the probability of the optimal path ending at position k where the last observation O_k has been emitted by state S_i. For the first position $k = 1$, $p_{i,\,k}$ is simply the probability of the observed state to be generated by each of the possible states multiplied by the prior probability of each state:

$$p_{i,1} = P(O_1|S_i)P(S_i)$$

Given $p_{i,k}$ for each i, we can calculate $p_{i,\,k+1}$. The probability $p_{i,\,k+1}$ is the probability of the most likely path of $k + 1$ hidden states assuming that the $k + 1$ observation was generated by state i. If the probability of the sequence through the optimal path up until the prior position is known, then this probability is easy to calculate:

$$p_{i,\,k+1} = P(O_{k+1}|S_i)P(S_i|S_j)p_k^{'}$$

where S_j is the prior state in the optimal path and $p_k^{'}$ is the probability of that optimal path up to the prior position. The probability is just the probability of the optimal path up until the prior state multiplied by the transition probability to the final state multiplied by the probability of the observation given that state. In practice, however, we do not know the prior state. But since we are searching for the most likely path we know the prior state must be the state that ultimately results in the highest probability. Since we have calculated $p_{i,k}$ for each possible prior state i, we can calculate the probability of the most likely path up to the next observation $k + 1$ where the state generating that observation is i:

$$p_{i,\,k+1} = P(O_{k+1}|S_i)\ \max_j\,(P(S_i|S_j)p_{j,\,k})$$

where j ranges over all of the different states. We note and store the state that is selected as the prior state. We can iterate through all of the observations, and calculate these probabilities. Once the entire sequence of n observations has been iterated through, then we select the state i that has the highest $p_{i,\,n}$ as the final state. The path of states to get to that state is the optimal path.

Now suppose instead that we wished to calculate the probability of a sequence of observations. A similar strategy can be used to calculate this probability over all possible states. This is known as the forward algorithm. It is similar to the Viterbi algorithm. At each step along the sequence we calculate $q_{i,k}$, which is the probability of the sequence up to position k over all possible states if the sequence ends in state i. So here we calculate the parameter

$$q_{i,\,k} = P(O_1 \ldots O_k,\ k\text{-th state} = S_i)$$

The initiation step is identical to the above:

$$q_{i,\,1} = P(O_1|S_i)P(S_i)$$

As we iterate through the sequence, we calculate:

$$q_{i,\,k+1} = P(O_{k+1}|S_i)\sum_j (P(S_i|S_j)q_{j,\,k})$$

Notice, the only difference from Viterbi is that instead of picking the best prior state, we sum over all prior states. The final probability of the sequence is the sum of all the probabilities for each of the different possible states at the end of the sequence, $q_{i,\,n}$.

Training hidden Markov models can be easy if a set of examples with known states is given. For example, many proteins with known crystal structures can be used to train the HMM depicted in Figure 2.26. The emission probability of the amino acids of each of the three states can be determined by empirical observation of the fraction of the amino acids observed in helices, loops, and beta sheets. Transition probabilities can be calculated as well by noting how frequently loops, helices, and sheets continue in the next amino acid in crystal structures and how often they transition to a new type of secondary structure. Similarly if a multiple alignment is available, the probabilities in Figure 2.25 can also be calculated by noting the frequency of deletions, insertions, and the probabilities of the amino acids in each of the different positions in the alignment.

However, often times, such multiple alignments are unavailable. All that is available is a collection of sequences. Under these circumstances the Baum–Welsh algorithm can be used to train HMMs. This is an expectation maximization algorithm in which we first assign random parameters to the model. Then we calculate the probability of the state for each sequence position. Finally we use those predicted probabilities to update the parameters.

For each position in a sequence of observed events we can calculate the probability of each state at a given position to be:

$$P(k\text{-th state} = S_i | O_1 \ldots O_n) = \frac{P(O_1 \ldots O_n,\ k\text{-th state} = S_i)}{P(O_1 \ldots O_n)}$$
$$= \frac{P(O_1 \ldots O_k,\ k\text{-th state} = S_i) P(O_{k+1} \ldots O_n | k\text{-th state} = S_i)}{P(O_1 \ldots O_n)}$$

The numerator can be split as above, since the probability of all of the observations after the k-th state depends only on that state, and does not depend at all on prior observations or states. The denominator of this value is taken from the forward algorithm; it is the sum of all $q_{i,n}$ terms over all states. The first term in the numerator is also from the forward algorithm; this is $q_{i,\,k}$. The second term can be calculated from the backward algorithm. The backward algorithm is a third dynamic programming algorithm in which we calculate terms:

$$r_{i,\,k} = P(O_{k+1} \ldots O_n | k\text{-th state} = S_i)$$

These terms are the probability of a sequence of observations occurring given that the system is in state i at position k. In this case we start from the back of the sequence. The initiation is that

$$r_{i,n} = 1$$

Then as we work backwards we calculate:

$$r_{i,\,k} = \sum_j P(O_{k+1} | S_j) P(S_j | S_i) r_{j,\,k+1}$$

So with the backwards algorithm combined with the forward algorithm we can calculate the probability of each state for each observation in a sequence.

To proceed with the Baum–Welsh algorithm we use the given collection of sequences and calculate the probability of each state at each position for each sequence:

$$P(k\text{-th state} = S_i | O_1 \ldots O_n) = \frac{q_{i,k} \cdot r_{i,k}}{\sum_j q_{j,n}}$$

Then we use these probabilities to re-estimate the parameters. So

$$P(O_j|S_i) \sim \frac{\sum_{k\text{-th obs} = O_j} P(k\text{-th state} = S_i)}{\sum_k P(k\text{-th state} = S_i)}$$

and

$$P(S_j|S_i) \sim \frac{\sum_k P(k+1\text{-th state} = S_j) P(k\text{-th state} = S_i)}{\sum_k P(k\text{-th state} = S_i)}$$

The Baum–Welsh algorithm guarantees that in every iteration the likelihood of the training data will increase. We repeat this procedure iteratively until the best possible parameters are obtained.

A very valuable application of the hidden Markov model is to actually do multiple alignments. We can use the Baum–Welsh approach to achieve this end. Typically we begin with a family of protein sequences and a hidden markov model with the structure illustrated in Figure 2.25. We can use the Baum–Welsh algorithm to iteratively fit the parameters of this HMM. As specified above, we first assign random parameters, then calculate the probability of each state for each position of each sequence, then update the parameters. Once the parameters are obtained, we have a hidden Markov model that describes this family of sequences. If the sequences are related and the parameters are appropriate, this could be a very useful model for the family. We can use the optimized HMM with the Viterbi algorithm to determine the hidden states of all of the sequences in the family. Each hidden state represents a position in the alignment. The position state determined for each protein sequence represents its position in the multiple alignment.

2.4 Gene expression profiling

High throughput measurement of the mRNA gene expression in cells is revolutionizing biology. Investigators are using technologies such as SAGE, oligonucleotide arrays, and spotted DNA microarrays to profile the gene expression of thousands of genes simultaneously. These studies are addressing a broad range of biological questions from human cancer to fruitfly development. Studies typically produce large gene expression data sets that contain measurements of thousands of genes under hundreds of conditions. There is a critical need to summarize this data and to pick out the important details. Otherwise, interpretation of the results is too difficult a task given the number and diversity of the genes. Many of the available analytical methods involve creating groups of genes or conditions that share properties in expression. Most of the commonly used strategies to accomplish this task utilize only the gene expression data. Other approaches include external information about genes and conditions. The most straightforward way to leverage external knowledge is to use binary statistical classification methods. In later chapters, this book will introduce strategies to analyze gene expression data sets with information from the scientific literature.

In order to illustrate how multiple methods can be applied to a single data set, we will explore a publicly available gene expression array data set consisting of 47 expression profiles of 4026 genes collected from lymphoma specimens (Alizadeh, Eisen et al. 2000). These profiles can be divided into two subtypes of lymphoma that have distinct clinical and molecular properties. We will apply some of the methods introduced in this chapter to this data set (Raychaudhuri, Sutphin et al. 2001).

The key concepts introduced in this section are described in the frame box. We commence this section with a brief introduction of experimental methods to measure gene expression. Then we describe metrics that can be used to calculate similarity and dissimilarity between gene expression profiles. We show how these metrics can be used to cluster gene expression profiles. The section closes with an introduction to classification methods, and their application to gene expression data.

1) Methods to measure gene expression
   a) Gene expression arrays
   b) Serial analysis of gene expression
2) Gene expression profile metrics
3) Clustering (unsupervised machine learning)
   a) K-means clustering
   b) Self-organizing maps
   c) Hierarchical clustering
4) Principal components analysis
5) Classification
   a) Nearest neighbor classification
   b) Linear discriminant analysis

2.4.1 Measuring gene expression with arrays

Gene expression technology permits the rapid assaying of mRNA quantities within individual cells. The rate of synthesis of a gene's protein product is approximately proportional to the amount of corresponding mRNA present within the cell. Most gene expression arrays do not measure absolute mRNA quantities; they measure the relative mRNA expressed within a cell subjected to an experimental condition compared to one subjected to a control condition. As depicted in Plate 2.6, the population of cells is divided in a typical expression profiling experiment; one half is subjected to some experimental environment while the other is subjected to a control environment. Gene expression arrays are then used to determine relative induction of genes within the experimental condition. These conditions may be different time points during a biological process, such as the yeast cell cycle (Cho, Campbell et al. 1998; Spellman, Sherlock et al. 1998) and drosophila development (White, Rifkin et al. 1999); direct genetic manipulations on a population of cells such as gene deletions (Hughes, Marton et al. 2000); or they can be different tissue samples with some common phenotype (such tissue type or malignancy) (Alizadeh, Eisen et al. 2000).

One popular gene expression array fabrication protocol involves spotting cDNA for specific genes at specified positions on a glass slide; each cDNA spot binds mRNA expressed from a particular gene (Plate 2.6) (Schena, Shalon et al. 1995). Another protocol involves synthesis of short oligonucleotide sequences onto specified positions on a solid substrate directly, using specific photolithographic techniques; each oligonucleotide spot binds specifically to mRNA expressed from specific genes (Chee, Yang et al. 1996). Nylon gene arrays have also been described (Chen, Wu et al. 1998). To measure mRNA quantities, mRNA is first harvested from cells. The mRNA is used as a template to synthesize proportional amounts of chemically labeled cDNA; typically the cDNA from the control and experimental conditions are chemically labeled with dyes that fluoresce at different wavelengths. All of the labeled cDNA is then hybridized to the gene array. At each spot the fluorescent intensity at the two wavelengths is measured; the ratio of intensities is reported as the relative expression of the corresponding gene. Most of the analytical methods described subsequently are applied to the log of these ratios.

Besides measuring gene expression, gene arrays have found other genomics applications as well. They have also been used to identify gene deletions (Behr, Wilson et al. 1999), gene duplications (Pollack, Perou et al. 1999), transposon locations (Raychaudhuri, Stuart et al. 2000), and single nucleotide polymorphisms (Halushka, Fan et al. 1999).

2.4.2 Measuring gene expression by sequencing and counting transcripts

An alternative strategy to using gene arrays to assay gene expression is Serial Analysis of Gene Expression, or SAGE (see Plate 2.7) (Velculescu, Zhang et al. 1995). SAGE is a considerably more intensive assay, but it permits quantitative assaying of large numbers of transcripts. SAGE assumes that large genetic sequences can be recognized by small 11–21 nucleotide "tags". SAGE works by obtaining these short tag sequences, concatenating them, and efficiently sequencing them. In addition SAGE assumes that the number of times these short sequences are observed among expressed sequences represents the level of expression.

The SAGE assay begins by isolating mRNA transcripts; the mRNA is used as a template to synthesize proportional amounts of cDNA with reverse transcriptase and a poly-T primer. Restriction enzymes are then used to splice out the short tags from the mRNA. Typically, a restriction enzyme recognizes short nucleotide sequences and cleaves at that site; it also leaves a short single-stranded overhang that is capable of binding its complement. The enzyme used to cleave the cDNA is called the "anchoring enzyme". Beads are used to bind to the poly-A tail of the shorter cleaved cDNA fragments. Beads and the attached segments are divided into two groups. The overhang from each set is used to bind and attach a sequence containing a second enzyme recognition site and one of two primer sites. The second enzyme site is recognized by a "tagging enzyme", which is a special restriction enzyme that cuts at a defined distance up to 20 base pairs downstream from the recognition site. After cleavage with the tagging enzyme, the remaining short sequence segments contain the tags from the sequence. These are the short stretches of the cDNA sequence downstream from the anchoring enzyme site that remain after cleavage with the tagging enzyme. The short sequence segments from the two sets are joined together to create ditags; the primers on the ends of the ditags are used for PCR amplification. The anchoring enzyme is then used again to cleave the primer sites, and the ditags are concatenated into larger sequences. These sequences are then sequenced, and the tags are identified. The tag concatenation permits rapid serial tag sequencing. The tag sequences that are obtained can be directly compared to known sequences of genes. The number of tags corresponding to the transcript of a particular gene gives a very good estimate of the expression of the gene.

SAGE is a very powerful, but labor intensive method. Since its introduction in 1995, SAGE has been applied widely to investigate many biological problems including gene expression profiling in many different human cells from different organs, cancer cells, and cells from other organisms as well (Hermeking 2003; Tuteja and Tuteja 2004; Tuteja and Tuteja 2004). Many of the analytical strategies discussed below can be used effectively on matrices of SAGE transcript counts.

2.4.3 Expression array analysis

Gene expression data sets may include measurements for thousands of genes across hundreds of conditions. Most expression analysis methods analyze data as a collection of either genes or conditions, each with a series of associated expression measurements called a "profile". If we imagine a two-dimensional array of measurements in which the rows are the measurements associated with individual genes and the columns are the measurements associated with conditions, the profile is the list of measurements along each row or column. "Features" are the individual expression measurements within each profile (see Figure 2.28). Depending on the analytical task, some features are more valuable than others; in many situations, focusing on a subset of the features improves results.

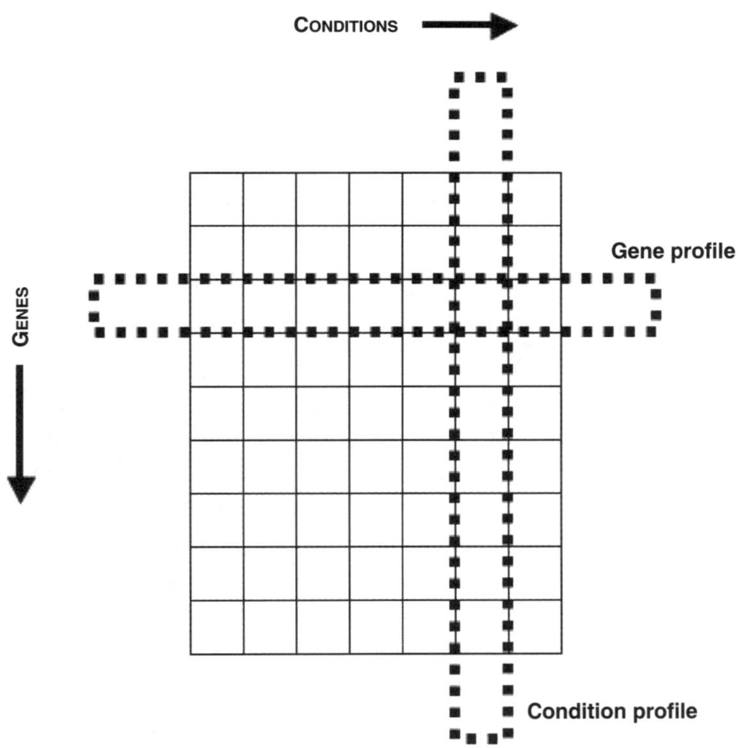

Figure 2.28 *Matrix of gene expression data.* Gene expression data can be organized into a matrix for easy analysis. Each row represents an individual gene, and each column represents a specific condition. A row therefore is a gene expression profile, while a column is a profile for a condition across all genes.

Because there is a symmetric relationship between genes and conditions, any data set can be analyzed in two ways. For example, we can interpret the lymphoma data set as 47 cancer profiles with 4026 available features, where each feature is the expression for a particular gene. In this case, we are trying to analyze the different cancer cases to understand their similarities and differences. Alternatively, we can analyze the genes, by interpreting the data as 4026 gene profiles with 47 available features, where each feature is the expression within a particular cancer specimen. Most analyses that can be performed on genes can also be performed in a symmetric manner on conditions. In the rest of the section we will talk mostly about analyzing genes; but these methods apply equally to genes and conditions.

Analytical algorithms that summarize the data have been applied to many data sets. Most popular are clustering (or unsupervised machine learning) algorithms that group together elements (such as genes or conditions) of these large data sets. Dimensional reduction approaches are also commonly applied; they reduce the number of features so that redundant ones that are very similar to others are removed or combined.

Other analytical methods include external information about genes and conditions into the analysis. Classification (or supervised machine learning) methods classify unknown cases by comparison to labeled training examples. These methods offer an avenue to include binary labels to genes (e.g. whether or not a gene has a specific function) or conditions. The limitation of these methods, however, is that they require the user to determine the relevant labels beforehand and provide labeled examples. Frequently, the relevant labels are not known in advance, or the known labels are inaccurate or incomplete.

2.4.4 Unsupervised grouping: clustering

Clustering methods help to simplify data sets by grouping profiles and decomposing the results into easier to interpret underlying "programs" of expression. These methods make the initial interpretation of expression data facile. However, the majority of clusters tend to be spurious and have little biological meaning and these can be difficult to interpret.

Some of the gene expression clusters have biological significance; it is these clusters that are the most important. Theoretically if genes have similar expression over a large number of conditions, it is possible that they may be regulated by similar mechanisms and they may have similar function. The value of an automated grouping method was apparent in one of the first large-scale gene expression studies (DeRisi, Iyer et al. 1997). Investigators manually identified five distinct subsets of genes with similar biological function that were coherently expressed in a yeast time series where media metabolites were altered. In early demonstrations of automated clustering algorithms, gene clusters derived from large expression studies on many conditions corresponded to certain particular biological functions; this offered promise for annotating uncharacterized genes and understanding the control of gene regulation (Eisen, Spellman et al. 1998; Michaels, Carr et al. 1998). Many have used gene expression clusters as a starting point for gene annotation and for understanding gene regulation.

Clustering algorithms group similar profiles together based on a distance metric—a formula for calculating the similarity between two profiles (Ripley 1996). There are many ways to express the distance between two numerical vectors. Many clustering algorithms are based on the statistical correlation coefficient (ranging from −1 to +1):

$$D(x, y) = 1 - \frac{xy'}{\| x \| \| y \|}$$

where x and y are vectors containing the expression values for two different genes. The distance is actually one minus the correlation coefficient to insure that two identical profiles have a distance of zero. Others use the Euclidean distance, the square root of the sum of the squared differences in corresponding features values:

$$D(x, y) = \| x - y \|$$

Investigators have devised new robust and efficient clustering methods specifically for gene expression studies recently (Altman and Raychaudhuri 2001). More appropriate metrics for expression studies that account for the sequential nature of time series measurements or eliminate outlier data have been proposed (Heyer, Kruglyak et al. 1999; Aach and Church 2001). Other groups have suggested methods for measuring cluster stability (Kerr and Churchill 2001; Ben-Hur, Elisseeff et al. 2002). One group investigated whether dimensional reduction techniques affected clustering (Yeung and Ruzzo 2001). In practice, however, the clustering methods most commonly applied to gene expression data are hierarchical clustering, self-organizing maps, and *k*-means clustering (Sherlock 2000).

The results of clustering can be very sensitive to the features that are used to compute the distance metric. Features are usually weighed equally and the effects of the relevant features can be masked by less relevant ones. For example, for a study of the response of cancer profiles to a pharmacological agent, a feature set including the entire genomic expression profile might not be appropriate because the response might depend only on a handful of target and transport genes, and inclusion of thousands of other genes might make similarities or differences difficult to extract from the noise of the irrelevant genes (Ross, Scherf et al. 2000).

2.4.5 *K*-means clustering

One of the simplest clustering methods is *k*-means clustering. It is very easy to implement.

K-means clustering requires a parameter *k*, the number of expected clusters. Correct selection of *k* can dramatically affect the final clustering results and unfortunately it is often difficult to know *a priori* what an appropriate choice for *k* is.

Initially *k* cluster centers, $c_1, \ldots, c_k$, are randomly selected expression profiles taken from the data set. In each iteration of the algorithm, the distances between each of the genes and the *k* centers are calculated using the pre-selected distance metric; genes are then assigned to the cluster whose center they are nearest to. For each gene x_j and each center c_i:

$$d_{j,\,i} = D(x_j,\, c_i)$$
$$\text{cluster}(x_j) = \arg\min_i\, (d_{j,\,i})$$

After the genes have been assigned to clusters, the cluster centers are recomputed by taking the average of the genes assigned to the cluster. In the subsequent iteration, genes are again assigned to the cluster whose center they are nearest to and then the centers are recalculated; this process is repeated until the algorithm converges. Unfortunately the algorithm converges to a local minimum, and is very sensitive to the initial random selection of starting centers.

We grouped the lymphoma cases in our test data set with *k*-means clustering. Such a calculation could be used to search for cancer subtypes—perhaps having unique biological or clinical properties. We used *k*-means clustering algorithm to cluster the lymphoma samples. To simplify the analysis, we did not use all 4026 genes as features, but instead used a subset of 148 that are expressed specifically in the germinal-center cell populations. The original investigators used the same subset in their cluster analysis (Alizadeh, Eisen et al. 2000). *K*-means clustering method with a Euclidean distance metric grouped the lymphoma cases into two clusters (Plate 2.8).

For this data set, the two clusters had different phenotypic properties. One cluster is composed of clinical cases with a poorer prognosis on average, the "activated" subtype. The other group, the "germinal center" subtype, specifically expressed these 148 genes. The success of this approach hinged on selecting an informative subset of features, the 148 germinal-center specific genes.

2.4.6 Self-organizing maps

Instead of simply partitioning data into disjoint clusters, self-organizing maps organize the clusters into a "map" where similar clusters are close to each other (Tamayo, Slonim et al. 1999). The number and topological configuration of the clusters are pre-specified. The method is similar to k-means clustering except that cluster centers are recalculated during each iteration based on the profiles within the cluster itself as well as the profiles in adjacent clusters. Over many iterations the clusters conform to the pre-specified topology. That is clusters that are near to each other in the predefined topology will contain genes that are similar to each other. This offers the user an advantage, particularly when dealing with large numbers of clusters; the algorithm organizes the clusters in a coherent fashion.

The user first defines a topology between the clusters. See Figure 2.29 for an example. Then cluster centers, $c_1, \ldots, c_k$, are assigned to be random profiles taken from the data set. At each iteration, a gene is selected randomly and the cluster center that is closest to it is identified. Say gene expression profile x is assigned closest to c_i and is therefore assigned to be a member of the i-th cluster. Then the cluster centers are updated with the following equation:

$$c_k = c_k + f_n(i, k)\mu(x - c_k)$$

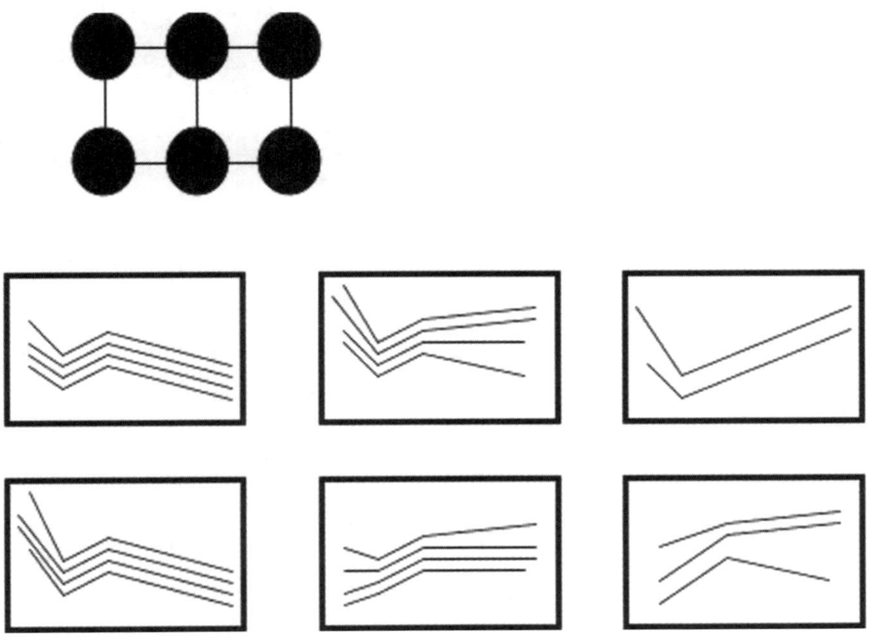

Figure 2.29 *Self-organizing map*. In self organizing maps a topology between clusters is predefined. In this schematic the clusters are arranged in a 2 × 3 grid. After the algorithm is run, expression profiles are organized into clusters. The profiles are most similar within the cluster. Clusters that are near to each other in the predefined topology are relatively similar to each other.

where μ is a small parameter that modulates the rate at which the centers are adjusted and f_n is the neighbor function that is inversely proportional to distance between clusters i and k in the pre-defined topology. So the cluster centers are recalculated at each step, and the amount they are adjusted is related to how close they are to the cluster that the selected gene is assigned to. This process is repeated until the clusters converge. The fact that expression profiles affect the center of their own cluster and also nearby clusters insures that clusters that are adjacent to each other in the pre-defined topology are similar to each other.

As a practical example we apply self-organizing maps to a yeast gene expression data set with 79 conditions measured on 2467 genes (Eisen, Spellman et al. 1998). Some of these clusters correlate with biological function. Looking at the average profile for each cluster, it is apparent that topologically close clusters are similar to each other (Figure 2.30).

2.4.7 Hierarchical clustering

Hierarchical clustering was the first clustering algorithm applied to high throughput gene expression data (Eisen, Spellman et al. 1998). The

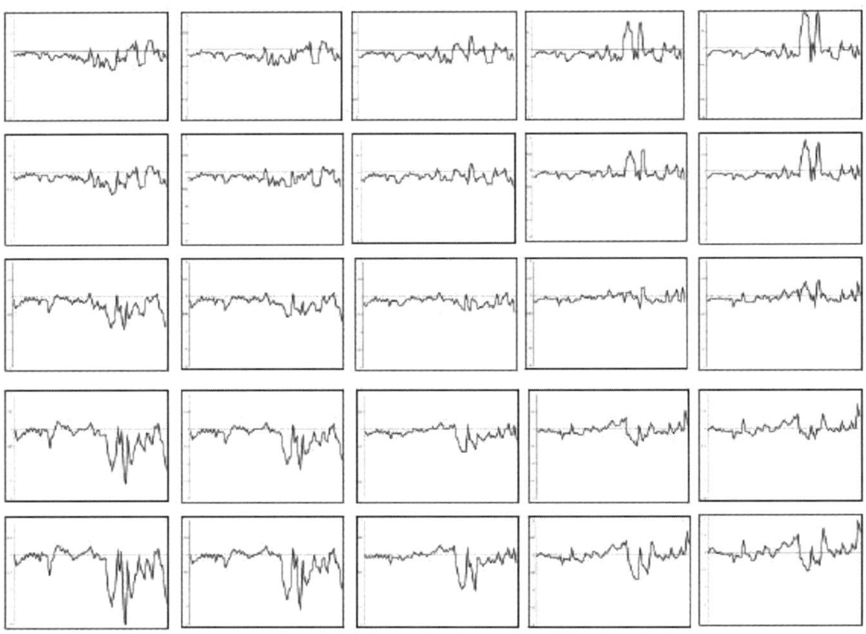

Figure 2.30 *Self-organizing map of yeast gene expression data.* 2467 genes were clustered over 79 conditions. Clusters were arranged in a 5 × 5 grid. Each graph represents a cluster of genes. The graphs are arranged according to the self-organizing map topology. In each of the graphs the average gene expression profile of all of the genes in that cluster is displayed over the conditions. Similarity between adjacent clusters is apparent.

algorithm analyzes the data and presents genes (or conditions) in the form of a dendrogram, or tree, based on gene expression similarity. The closer two genes are placed together in the dendrogram the more related they are in terms of gene expression. This clustering method predates gene expression analysis, and was a favorite approach to clustering gene and protein sequences. Hierarchical clustering strategies suffer because the decision about where to create branches and in what order the branches should be arranged can be arbitrary. In practice biologists often use their knowledge about genes to determine whether they appear related to each other and draw appropriate boundaries in hierarchical clusters manually. One of the reasons why many prefer hierarchical clustering is that it offers the user some flexibility to draw the cluster boundaries.

Here we describe agglomerative hierarchical clustering. This form of clustering starts at the twigs of the tree and works its way up to the trunk (see Figure 2.31). Say hierarchical clustering is applied to a data set that has

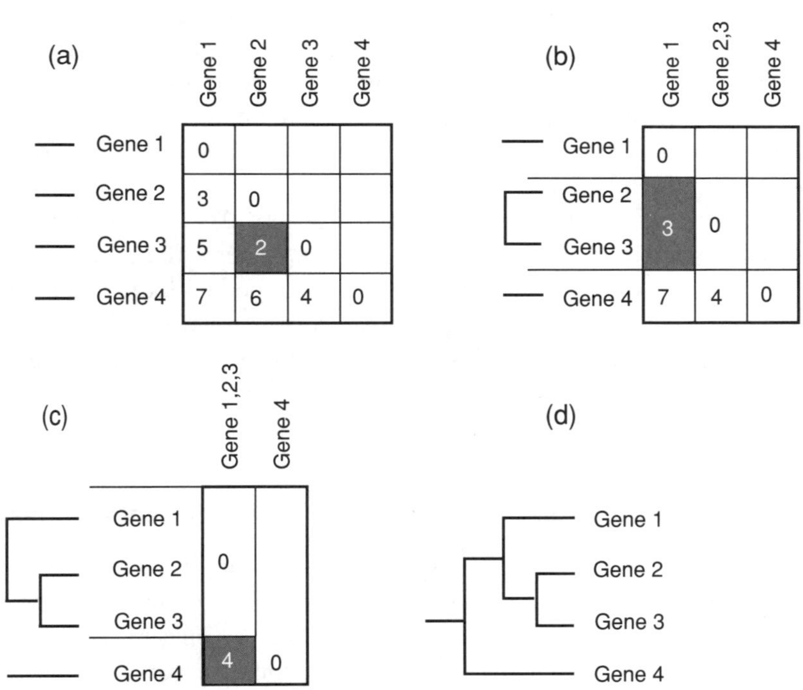

Figure 2.31 *Agglomerative hierarchical clustering.* To create a hierarchical clustering of genes we begin by calculating a distance matrix, D, between all of the genes. (a) The first step involves identifying the nearest pair of genes. In this example Gene 2 and Gene 3 are the most similar to each other. (b) The two nearest genes are merged. The distances between this merged entity and the remaining genes are recalculated. In this case the minimum distance is used. A new distance matrix is created. (c) Again the nearest two entities in the redefined distance matrix are identified by noting the smallest value in the distance metric. These two entities are then combined. (d) This process is repeated iteratively until only one merged entity (the root) remains.

gene expression profiles for n genes. Each of the n genes is initially considered an individual cluster. The task is then to merge them into larger and larger clusters, until they have all been combined into a single cluster.

First, hierarchical clustering calculates the $n(n-1)/2$ pairwise distances between each of the n clusters of single genes using one of the aforementioned distance metrics. These distances are all stored in a matrix D. Next we search for the smallest off-diagonal distance in D. The two most similar non-identical clusters are then merged into a new cluster; this defines the first branchpoint in the tree. Let us assume that these two profiles are i and j. We now recalculate the matrix D; since gene i and gene j have been merged, we remove the rows and columns corresponding to i and j in D. We add another row and column for the (i, j) cluster. So we have replaced two clusters containing only a single gene each with a single cluster containing two genes.

To update D, there are three options to calculate the distance between the new (i,j) cluster and the other remaining genes. The first is to calculate an average distance from the original distances to the individual genes (average linkage). Another option is to assume the distance between a gene and the new cluster is the greatest distance between that gene profile and all of the constituent gene profiles in the cluster (complete linkage). The third option is to assume that it is the least distance between that gene and the constituent genes (single linkage).

Once D has been updated, we repeat the process. Again we search for the smallest distance in D. We combine those two clusters, and update D. This time it is possible that a cluster of one gene is grouped with the larger cluster of two genes to form an even larger cluster of three genes. This process is repeated until all of the genes have been merged into a single cluster. The sequence in which the merges occurred determines the structure of the cluster tree.

2.4.8 Dimension reduction with principal components analysis

Like clustering algorithms, dimensional reduction algorithms also reduce the complexity of the data. Application of dimension reduction methods to gene array data is an alternative to clustering of genes (Raychaudhuri, Stuart et al. 2000). Like clustering methods, dimension reduction methods do not include any outside information besides the expression data itself. Dimension reduction involves removing or consolidating features in the data set. Features are removed because they do not provide any significant incremental information, and because they can confuse the analysis or make it unnecessarily complex. For example, a time series experiment may sample data more finely than necessary, and so many of the conditions are intercorrelated and do not offer additional information about the genes.

Instead, we would choose a subset of conditions that contains "independent" information. Dimension reduction can make the outliers and clusters in a data set apparent, and can also reduce the noise in the data set. It can suffer, however, by throwing away important but weak signals in the data.

Microarray data sets are sufficiently large that dimension reduction can help algorithms run more quickly, and can also make the results of an analysis easier to understand. Dimension reduction can be accomplished with a number of methods, including principal components analysis, singular value decomposition, independent components analysis, and others.

Principal component analysis (PCA) automatically detects redundancies in the data and defines a new (smaller) set of hybrid features, or components, that are guaranteed not to be redundant. The hybrid features, or principal components, are composites of the original features, chosen to provide separate information about the genes or conditions. Each principal component is a normalized linear combination of the original variables. These components together account for as much of the variance in the original n variables as possible while remaining mutually uncorrelated and orthogonal.

To compute the principal components for a dataset of m genes and n conditions, we first center the data, so that for each condition the mean expression is zero. Then we calculate the covariance matrix. The n eigenvalues and their corresponding eigenvectors are calculated from the $n \times n$ covariance matrix of conditions. Each eigenvector defines a principal component. A component can be viewed as a weighted sum of the conditions, where the coefficients of the eigenvectors are the weights. The projection of gene i along the axis defined by the $j^{-\text{th}}$ principal component is:

$$a_{ij}^{\text{PCA}} = \sum_{t=1}^{n} a_{it} v_{tj}$$

where v_{tj} is the $t^{-\text{th}}$ coefficient for the $j^{-\text{th}}$ principal component, and a_{it} is the expression measurement for gene i under the $t^{-\text{th}}$ condition. A^{PCA} is the data in terms of principal components. Since V is an orthonormal matrix of eigenvectors, A^{PCA} is a rotation of the data from the original space of observations to a new space with principal component axes.

The variance accounted for by each of the components is its associated eigenvalue; it is the variance of a component over all genes. Consequently, the eigenvectors with large eigenvalues are the ones that contain most of the information; eigenvectors with small eigenvalues are uninformative. We assume the components with low variance have little information and we eliminate them. Determining the true dimensionality of the data and the number of components to eliminate is often *ad hoc* and many heuristics exist.

We applied PCA to the test lymphoma data (Figure 2.32). In this case, we have reduced 47 lymphoma data points in 148-dimensional space (47 cancer cases associated with expression measurements for 148 germinal-center genes) to just two dimensions; each cancer profile is plotted as a 2D point in the graphic. Each spot in the graphic is shaded to match its corresponding cluster in Plate 2.8; the two subtypes are clearly separated in the reduced two-dimensional component plot. It is sometimes possible to interpret the new features biologically. For this data set, the first dimension may be a measure of average overall expression of germinal center specific genes.

2.4.9 Combining expression data with external information: supervised machine learning

Supervised classification approaches offer the most straightforward possibility of incorporating outside knowledge. Given a set of known cases, classification algorithms allow the possibility of determining whether unseen cases are similar to the given cases and therefore likely from the same class. The selection of the known cases is where the external information is injected into the analysis. These approaches, therefore, require a set of examples of expression profiles that are labeled with some phenotype or categorization. The rules that are devised from these examples are used to predict properties of unseen expression profiles. One advantage of

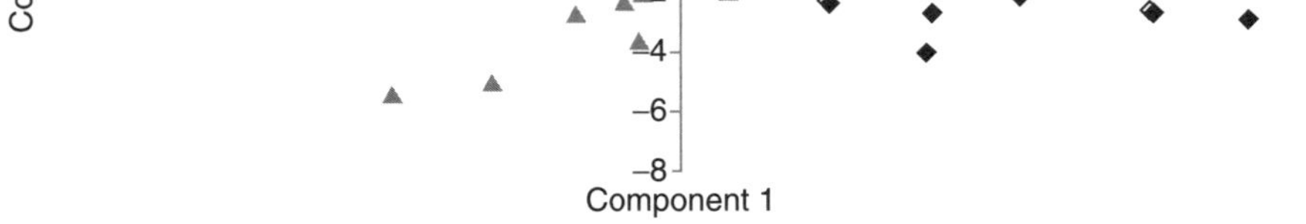

Figure 2.32 *Visualization of 148-dimensional lymphoma data in two dimensions using principal component analysis.* Principal component analysis (PCA) applied to the lymphoma expression profiles over the 148 germinal-center specific genes makes it possible to visualize the data in two-dimensional space. Approximately 45% of the total variance is contained in the first two dimensions. Each point in the figure represents a specific cancer profile. It is plotted in an expression space represented by two components. The cases from the germinal-center subtype are dark diamonds, and the activated subtype cases are lighter triangles. The clusters are well separated in this space.

supervised machine learning techniques is that they usually place differential weights on the features based on their utility in distinguishing between different categories.

For example, one application of classification algorithms is in predicting the function of a gene by comparison of its expression profile to those of well studied genes. Another application of classification algorithms is disease diagnosis based on the gene expression profile of a pathologic specimen taken from a patient's biopsy.

Typical use of classification approaches requires the selection of a positive and negative training set. The training sets contain the known cases. The positive set contains examples that belong to the class, such as genes with a particular function. The negative set contains examples of cases that do not belong to the class, such as genes that specifically are confirmed not to have that same function.

Examples of such methods include logistic regression, nearest neighbor classification, neural networks, and linear discriminant analysis (LDA). Logistic regression uses the feature values for different groups to estimate the parameters of a predictor function (a linear log-likelihood model) to best account for the known classified cases (Ripley 1996). Neural networks use a set of known examples to create a multi-layered computational network that produces a prediction of the category for each unknown case. We review linear discriminant analysis and nearest neighbor classification in greater detail below.

2.4.10 Nearest neighbor classification

Nearest neighbor classification schemes are some of the easiest to implement and understand, and frequently they are very effective methods. Given a previously unseen test case, nearest neighbor classification determines whether or not it belongs to the class by identifying the most similar cases. If many of these similar cases belong to the class, then it is assumed that the test case belongs to the class as well. So first the distance using a pre-selected metric between the test case, x, and each of the training examples, x_i, is obtained:

$$d_i = D(x, x_i)$$

Then, the closest k training examples are identified. If more than $k/2$ training examples are positive examples, we predict that the test case belongs to the class represented by the positive training set.

2.4.11 Linear discriminant analysis

Linear discriminant analysis uses the labeled examples from each set of classified cases to estimate a probability distribution for the values of the features in that class (see Figure 2.33). Given a new example, it uses the distributions to determine the most likely class and assigns the example to it.

The basic assumption of LDA is that the positive and negative training examples can be modeled with normal distributions. The probability density function of a multivariate normal distribution is:

$$F(x) = \frac{e^{-\frac{1}{2}(x-\mu)' S^{-1}(x-\mu)}}{\sqrt{(2\pi)^n |S|}}$$

The two critical parameters are the mean on which the distribution is centered, μ, and the covariance matrix, S, that determines the shape of the distribution. Let's assume that our training examples are in a matrix, X, where each row represents a gene's expression profile. The mean for the

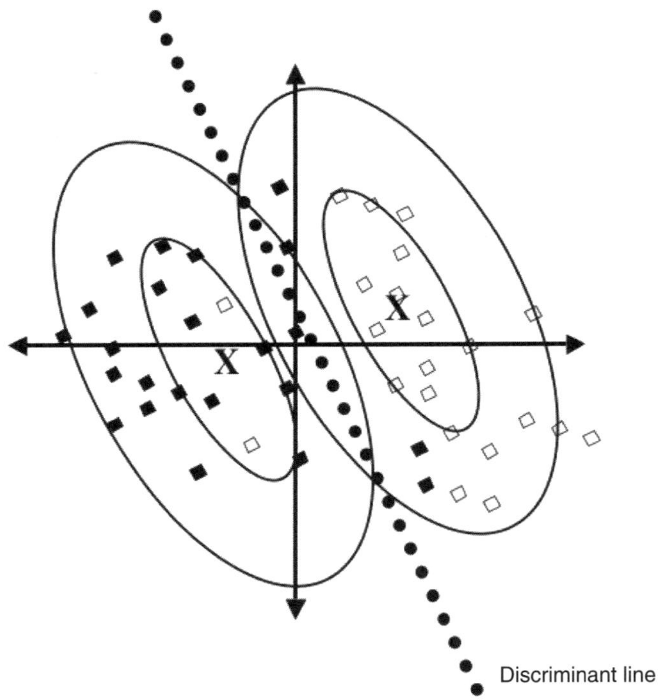

Figure 2.33 *Linear discriminant analysis.* This is an example of a discriminant line separating positive training cases (white squares) from negative training cases (dark squares). First for each of those sets of data a separate mean is calculated, indicated by the "X". A pooled covariance matrix is calculated for both sets of data separately. The pooled covariance matrix and the means are used to define two normal distributions: one that models the positive cases and one that models the negative cases. The ovals represent the normal distributions. The distributions are assumed to have the same covariance matrix – so they have a similar shape. The discriminant line, indicated here with a dotted line, is defined as the collection of points where the density of the positive distribution is equal to the density of the negative distribution.

positive training set and negative training set are calculated separately; we will denote them as μ^+ and μ^-, respectively. A common covariance matrix is assumed for both distributions. This pooled covariance matrix is the average of the covariance matrix of the positive training examples, S^+, and the covariance of the negative training examples, S^-:

$$S = \frac{1}{2}(S^+ + S^-)$$

Given an unseen test case, x, we can calculate the log of the ratio of the probability of x assuming that it was generated by the positive model to the probability of x assuming that it was generated by the negative model. This is the log likelihood that x is a positive case:

$$\log\left(\frac{p(x|+)}{p(x|-)}\right) = \log\left(\frac{F_+(x)}{F_-(x)}\right)$$
$$= -\frac{1}{2}(x - \mu^+)' S^{-1}(x - \mu^+) + \frac{1}{2}(x - \mu^-)' S^{-1}(x - \mu^-)$$

where F_+ is the normal distribution characterizing the positive training examples, and F_- is the normal distribution characterizing the negative training examples. Since we have assumed they have identical covariance matrices, we can further simplify to:

$$\log\left(\frac{p(x|+)}{p(x|-)}\right) = -\frac{1}{2}(\mu^+ - \mu^-)' S^{-1}(\mu^+ + \mu^-) + x' S^{-1}(\mu^+ - \mu^-)$$

If the value of the log likelihood is greater than zero, then we assume that the test case, x, is consistent with the positive set and is classified accordingly, otherwise we classify the test case with the negative set.

Supervised grouping methods can be applied to the lymphoma data set. For example, suppose the biological expertise required to pick out the 148 germinal center genes (as we did in the unsupervised grouping illustration) is not available. Instead, we would be faced with a data set consisting of expression measurements for 4026 genes. Suppose, however, that we know that there are two different clinical presentations of the disease. In particular, we can partition the data into two sets—those that fall into a less malignant group and those that fell into the more malignant group. To demonstrate supervised grouping, we select ten clear-cut cases of each—ten very aggressive lymphomas and ten very benign ones. Then use LDA to predict accurately the prognosis for the remaining 27 unknown cases (Plate 2.9).

The results of applying LDA to other gene expression classification tasks are described in Table 2.4; these tasks include gene function assignment and

Table 2.4 *Application of LDA supervised classification to diverse tasks.* The first two tasks are classifying human cancers (condition) expression profiles based on genome wide expression assays of pathologic specimens. The second two tasks are classifying the type of function or regulation of genes based on expression measurements. Each row represents a publicly available data set in which there are two predefined classes (arbitrarily designated as either positive or negative). The column "Gold standard reference" contains the primary resources where the data categorizations were obtained (and/or the original raw data). The columns "Positive set" and "Negative set" contain a description of the two classification categories and the number of examples for each (the sets are labeled positive and negative arbitrarily). The "Features" column states the number and type of features used for classification. For example the leukemia profile consists of measurements over 7129 genes. Alternatively, in cases where genes are being classified, its expression in each array experiment constitutes the features. The "Cross-validation accuracy" column is an estimate of the percentage of correctly classified examples on unseen data. We classified acute leukemia cases into the well-established clinical subtypes of myeloid and lymphoid leukemia based on gene expression in pathologic specimens with linear discriminant analysis (LDA). We classified cancerous and non-cancerous cell lines into diffuse non-B cell lymphoma (DLCL) and non-DLCL cell lines based on gene expression in pathologic specimens. We classified yeast genes into members and non-members of the ribosomal complex, as defined by the MIPS consortium based on gene expression in yeast under diverse conditions. We also distinguished between yeast genes regulated by the *mse* upstream promoter element versus those regulated by a *urs1* upstream promoter element; both promoters are critical to yeast sporulation. The data set consisted of a yeast sporulation time series as well as non-related cell-cycle and metabolic time series experiments.

Problem	Gold standard reference	Positive set N	Negative set N	Features	Cross-validation accuracy
1. Acute leukemia (Golub, Slonim et al. 1999)	Primary Data (Golub, Slonim et al. 1999)	Lymphoid 47	Myeloid 25	7129 genes	95.83%
2. Lymphoma (Alizadeh, Eisen et al. 2000)	Primary Data (Alizadeh, Eisen et al. 2000)	Diffuse large cell lymphoma 42	Non-DLCL 54	4026 genes	95.83%
3. Ribosomal genes (Eisen, Spellman et al. 1998)	MIPS catalogue (Mewes, Frishman et al. 2000)	Ribosomal genes 121	Other Genes 2346	79 arrays	99.23%
4. Sporulation promoters (DeRisi, Iyer et al. 1997; Chu, DeRisi et al. 1998; Spellman, Sherlock et al. 1998)	Reviews (Mitchell 1994; Chu and Herskowitz 1998)	Early genes (URS1 promoters) 13	Middle genes (MSE promoters) 23	103 arrays	97.20%

cancer subtype classification. Clearly, the success of supervised machine learning is dependent on whether high-quality labeled sets are provided. In the case of expression data, supervised grouping will certainly fail if the expression data does not explain the phenotype in question.

References

Aach, J. and G. M. Church (2001). "Aligning gene expression time series with time warping algorithms." *Bioinformatics* **17**(6): 495–508.

Alberts, B., D. Bray, et al. (1994). *Molecular Biology of the Cell*. New York, Garland Publishing.

Alizadeh, A. A., M. B. Eisen, et al. (2000). "Distinct types of diffuse large B-cell lymphoma identified by gene expression profiling." *Nature* **403**(6769): 503–11.

Altman, R. B. and S. Raychaudhuri (2001). "Whole-genome expression analysis: challenges beyond clustering." *Curr. Opin. Struct. Biol.* **11**(3): 340–7.

Altschul, S. F., W. Gish, et al. (1990). "Basic local alignment search tool." *J. Mol. Biol.* **215**(3): 403–10.

Altschul, S. F., T. L. Madden, et al. (1997). "Gapped BLAST and PSI-BLAST: a new generation of protein database search programs." *Nucleic Acids Res.* **25**(17): 3389–402.

Behr, M. A., M. A. Wilson, et al. (1999). "Comparative genomics of BCG vaccines by whole-genome DNA microarray." *Science* **284**(5419): 1520–3.

Ben-Hur, A., A. Elisseeff, et al. (2002). "A stability based method for discovering structure in clustered data." *Pac. Symp. Biocomput*: 6–17.

Chee, M., R. Yang, et al. (1996). "Accessing genetic information with high-density DNA arrays." *Science* **274**(5287): 610–4.

Chen, J. J., R. Wu, et al. (1998). "Profiling expression patterns and isolating differentially expressed genes by cDNA microarray system with colorimetry detection." *Genomics* **51**(3): 313–24.

Cho, R. J., M. J. Campbell, et al. (1998). "A genome-wide transcriptional analysis of the mitotic cell cycle." *Mol. Cell.* **2**(1): 65–73.

Chu, S., J. DeRisi, et al. (1998). "The transcriptional program of sporulation in budding yeast." *Science* **282**(5389): 699–705.

Chu, S. and I. Herskowitz (1998). "Gametogenesis in yeast is regulated by a transcriptional cascade dependent on Ndt80." *Mol. Cell.* **1**(5): 685–96.

DeRisi, J. L., V. R. Iyer, et al. (1997). "Exploring the metabolic and genetic control of gene expression on a genomic scale." *Science* **278**(5338): 680–6.

Durbin, R., S. Eddy, et al. (2003). *Biological Sequence Analysis*. Cambridge, Cambridge University Press.

Eisen, M. B., P. T. Spellman, et al. (1998). "Cluster analysis and display of genome-wide expression patterns." *Proc. Natl. Acad. Sci. U S A.* **95**(25): 14863–8.

Golub, T. R., D. K. Slonim, et al. (1999). "Molecular classification of cancer: class discovery and class prediction by gene expression monitoring." *Science* **286**(5439): 531–7.

Gotoh, O. (1982). "An improved algorithm for matching biological sequences." *J. Mol. Biol.* **162**(3): 705–8.

Halushka, M. K., J. B. Fan, et al. (1999). "Patterns of single-nucleotide polymorphisms in candidate genes for blood-pressure homeostasis." *Nat. Genet.* **22**(3): 239–47.

Hermeking, H. (2003). "Serial analysis of gene expression and cancer." *Curr. Opin. Oncol.* **15**(1): 44–9.

Heyer, L. J., S. Kruglyak, et al. (1999). "Exploring expression data: identification and analysis of coexpressed genes." *Genome. Res.* **9**(11): 1106–15.

Hughes, T. R., M. J. Marton, et al. (2000). "Functional discovery via a compendium of expression profiles." *Cell* **102**(1): 109–26.

Kerr, M. K. and G. A. Churchill (2001). "Bootstrapping cluster analysis: assessing the reliability of conclusions from microarray experiments." *Proc. Natl. Acad. Sci. USA.* **98**(16): 8961–5.

Krogh, A., M. Brown, et al. (1994). "Hidden Markov models in computational biology. Applications to protein modeling." *J. Mol. Biol.* **235**(5): 1501–31.

Mantovani, R. (1998). "A survey of 178 NF-Y binding CCAAT boxes." *Nucleic Acids Res.* **26**(5): 1135–43.

Mewes, H. W., D. Frishman, et al. (2000). "MIPS: a database for genomes and protein sequences." *Nucleic Acids Res.* **28**(1): 37–40.

Michaels, G. S., D. B. Carr, et al. (1998). "Cluster analysis and data visualization of large-scale gene expression data." *Pac. Symp. Biocomput*: 42–53.

Mitchell, A. P. (1994). "Control of meiotic gene expression in Saccharomyces cerevisiae." *Microbiol. Rev.* **58**(1): 56–70.

Needleman, S. B. and C. D. Wunsch (1970). "A general method applicable to the search for similarities in the amino acid sequence of two proteins." *J. Mol. Biol.* **48**(3): 443–53.

Nelson, D. L., A. L. Lehninger, et al. (2000). *Lehninger Principles of Biochemsitry*, Worth Publishing.

Pearson, W. R. (1990). "Rapid and Sensitive Sequence Comparison with FASTP and FASTA." *Methods in Enzymology* **183**: 63–98.

Pearson, W. R. and D. J. Lipman (1988). "Improved tools for biological sequence comparison." *Proc. Natl. Acad. Sci. U S A.* **85**(8): 2444–8.

Pollack, J. R., C. M. Perou, et al. (1999). "Genome-wide analysis of DNA copy-number changes using cDNA microarrays." *Nat. Genet.* **23**(1): 41–6.

Raychaudhuri, S., J. M. Stuart, et al. (2000). "Principal components analysis to summarize microarray experiments: application to sporulation time series." *Pac. Symp. Biocomput*: 455–66.

Raychaudhuri, S., J. M. Stuart, et al. (2000). "Pattern recognition of genomic features with microarrays: site typing of Mycobacterium tuberculosis strains." *Proc. Int. Conf. Intell. Syst. Mol. Biol.* **8**: 286–95.

Raychaudhuri, S., P. D. Sutphin, et al. (2001). "Basic microarray analysis: grouping and feature reduction." *Trends Biotechnol.* **19**(5): 189–93.

Ripley, B. D. (1996). *Pattern Recognition and Neural Networks*. New York, Cambridge University Press.

Ross, D. T., U. Scherf, et al. (2000). "Systematic variation in gene expression patterns in human cancer cell lines." *Nat. Genet.* **24**(3): 227–35.

Schena, M., D. Shalon, et al. (1995). "Quantitative monitoring of gene expression patterns with a complementary DNA microarray." *Science* **270**(5235): 467–70.

Sherlock, G. (2000). "Analysis of large-scale gene expression data." *Curr. Opin. Immunol.* **12**(2): 201–5.

Shi, H. and P. B. Moore (2000). "The crystal structure of yeast phenylalanine tRNA at 1.93 A resolution: a classic structure revisited." *Rna.* **6**(8): 1091–105.

Spellman, P. T., G. Sherlock, et al. (1998). "Comprehensive identification of cell cycle-regulated genes of the yeast Saccharomyces cerevisiae by microarray hybridization." *Mol. Biol. Cell.* **9**(12): 3273–97.

Stryer, L. (1995). *Biochemistry.* New York City, W.H. Freeman and Company.

Tamayo, P., D. Slonim, et al. (1999). "Interpreting patterns of gene expression with self-organizing maps: methods and application to hematopoietic differentiation." *Proc. Natl. Acad. Sci. USA.* **96**(6): 2907–12.

Thompson, J. D., D. G. Higgins, et al. (1994). "CLUSTAL W: improving the sensitivity of progressive multiple sequence alignment through sequence weighting, position-specific gap penalties and weight matrix choice." *Nucleic Acids Res.* **22**(22): 4673–80.

Tuteja, R. and N. Tuteja (2004). "Serial analysis of gene expression: Applications in human studies." *J. Biomed. Biotechnol.* **2004**(2): 113–120.

Tuteja, R. and N. Tuteja (2004). "Serial analysis of gene expression: Applications in malaria parasite, yeast, plant, and animal studies." *J. Biomed. Biotechnol.* **2004**(2): 106–112.

Velculescu, V. E., L. Zhang, et al. (1995). "Serial analysis of gene expression." *Science* **270**(5235): 484–7.

White, K. P., S. A. Rifkin, et al. (1999). "Microarray analysis of Drosophila development during metamorphosis." *Science* **286**(5447): 2179–84.

Williams, J. C., J. P. Zeelen, et al. (1999). "Structural and mutagenesis studies of leishmania triosephosphate isomerase: a point mutation can convert a mesophilic enzyme into a superstable enzyme without losing catalytic power." *Protein Eng.* **12**(3): 243–50.

Yeung, K. Y. and W. L. Ruzzo (2001). "Principal component analysis for clustering gene expression data." *Bioinformatics* **17**(9): 763–74.

3 Textual profiles of genes

Using algorithms to analyze natural language text is a challenging task. Recent advances in algorithms, and increased availability of computational power and online text has resulted in incremental progress in text analysis (Rosenfeld 2000). For certain specific applications natural language processing algorithms can rival human performance. Even the simplest algorithms and approaches can glean information from the text and do it at a rate much faster than humans. In the case of functional genomics, where an individual assay might include thousands of genes, and tens of thousands of documents pertinent to those genes, the speed of text mining approaches offers a great advantage to investigators trying to understand the data. In this chapter, we will focus on techniques to convert text into simple numerical vectors to facilitate computation. Then we will go on to discuss how these vectors can be combined into textual profiles for genes; these profiles offer additional biologically meaningful information that can complement available genomics data sets.

The previous chapter introduced methods to analyze gene expression data and sequence data. The focus of many analytical methods was comparing and grouping genes by similarity. Some sequence analysis methods like dynamic programming and BLAST offer opportunities to compare two sequences, while multiple sequence alignment and weight matrices provide a means to compare families of sequences. Similarly, gene expression array analysis approaches are mostly contingent on distance metrics that compare gene expression profiles to each other; clustering and classification algorithms provide a means to group similar genes. The primary goal of applying these methods was to transfer knowledge between similar genes.

We can think of the scientific literature as yet another data type and define document similarity metrics. Algorithms that tap the knowledge locked in the scientific literature require sophisticated natural language processing approaches. On the other hand, assessing document similarity is a comparatively easier task. A measure of document similarity that corresponds to semantic similarity between documents can also be powerful. For example, we might conclude that two genes are related if documents that refer to them are semantically similar. We assess similarity by looking at the words that are used in the documents, and seeing if these words are the same or similar.

The concepts presented in this chapter are listed in the frame box. First we will introduce document vectors and the common metrics to measure similarity between them. Then we talk about strategies to remove the less semantically meaningful words and to also weight words according to their semantic value. We then talk about recasting document vectors in a reduced dimensional space using latent semantic indexing. We show how textual profiles for genes can be created. We demonstrate the utility of these textual profiles in the context of gene expression data, sequence data, and finding keywords that describe a gene. We then close with a discussion on strategies to query genes to identify genes with specific biological functions.

3.1 Representing documents as word vectors

One of the simplest and most effective representations for a document is the word vector model (Manning and Schutze 1999). A document can be converted into a word vector by simply counting the number of occurrences of each word. For a document d_j where each word i is present a_{ij} times:

$$d_j = \langle a_{1j}, a_{2j}, a_{3j}, \ldots, a_{Nj} \rangle$$

We demonstrate the conversion of a document to a word vector in Figure 3.1. Given a large collection of documents, we can construct a matrix, A, that represents the entire corpus of documents. Each column is a word vector for a specific document, and each row represents a particular word. Each entry a_{ij} corresponds to the number of times word i appears in

1) Document word vectors
2) Document vector distance metrics
3) Feature selection and weighting
   a) Zipf's law
   b) Removing stop words
   c) Stemming
   d) Word weighting
4) Latent semantic indexing: dimension reduction
5) Using reference indices to build gene textual profiles
6) Gene textual similarity correlates with sequence similarity
7) Gene textual similarity correlates with expression similarity
8) Keyword assignment to genes and groups of genes
9) Querying gene text for biological functions

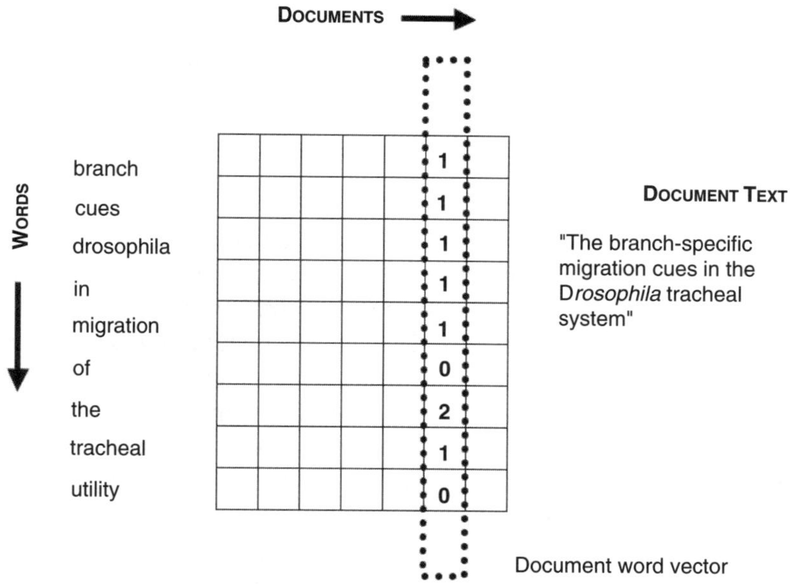

Figure 3.1 *Converting document text to a word vector.* To convert document text to a word vector, words in the textual document are counted. Those counts are entered in the vector. In many cases, multiple document vectors can be compiled together in a matrix. For a given document vector, the vast majority of words are present zero times, and the vector is quite sparse.

document j. We use the term document very broadly. Depending on the application, the document might be a whole full text journal article, an article abstract, a particular paragraph, a single sentence, or a collection of words in a particular phrase. Matrix A may have a very large number of rows, since the number of vocabulary words represented in the rows of A might be enormous. A typical vocabulary may include thousands of words. The matrix A may have a large number of columns as well, since most functional genomics corpora contain tens of thousands of documents. In addition, A is often a very sparse matrix. Most documents will only use a minority of the available vocabulary. For example a typical PubMed abstract may contain several hundred words. Since the vocabulary may include ten thousand words, only a handful of entries for the document's vector will be non-zero. Most vocabulary words will not occur among the mere several hundred words used in the text.

There are a few definitions with regards to this matrix that are worth noting. The a_{ij} matrix term is also known as the *term frequency*; it is the number of times a particular word shows up in a specific document. We will also refer to the term frequency of word i in document j as tf_{ij}. The document frequency, df_i, of a word i is the number of documents in the corpus that the term appears in. The collection frequency, cf_i, of a word i is the total number of times that a word appears in a corpus. Given any word, its collection frequency is the sum of its term frequencies over all documents:

$$\sum_j tf_{ij} = cf_i$$

The advantage of the word vector model is that standard matrix analysis approaches can now be applied to a corpus of documents. For example, many of the same strategies that we used to analyze the gene expression data matrices introduced in section 2.4 can be directly applied to document data in this form. In Plate 3.1 we illustrate with a hierarchical clustering of article abstracts about gene expression analysis (Altman and Raychaudhuri 2001). Here, we have clustered documents with *Cluster*, a hierarchical clustering software package available for gene expression analysis (Eisen, Spellman et al. 1998). Looking at the titles of these articles organized in the hierarchy, we see that related articles tend to cluster together. Other clustering strategies, dimensional reduction approaches, and classification methods can also be applied to these matrices.

In the subsequent sections we will also talk about ways to make more effective word vectors by removing semantically less valuable words, differentially weighting words, and applying dimensional reduction methods.

The word vector document model was originally exploited in information retrieval tasks, such as retrieving appropriate documents given a keyword query. These applications have become valuable in internet and literature database searches (Hersh 2003). While this approach is very convenient for computation, a significant amount of information is lost in this document representation. Basic sentence structure is completely obliterated, and it may be very difficult to piece together the meaning of a document from the word vector. For example, one very challenging aspect of this documentation is negations. A document that says a gene has a certain function and another document that says that the same gene definitively lacks that function have semantically opposite theses. But, from the point of view of word vectors, these documents look very similar. They may use the exact same words to describe the gene, the function, and the assays used to derive the ultimate conclusion. Consequently word vector similarity can also be confounding for certain applications.

3.2 Metrics to compare documents

The advantage of vector formulations is that assessing document similarity becomes a relatively easy task. There are many approaches, each of which may be appropriate for a given situation (Manning and Schutze 1999). The simplest metrics measure distances between binary vectors, while more sophisticated metrics take into account the actual term frequencies. Binary vectors can be formulated from any of the above vector formulations by replacing non-zero term frequencies with one; this indicates only the presence or absence of a single word in a document vector.

The *matching coefficient* is the simplest similarity measure between binary vectors; it counts the number of dimensions that are non-zero for two documents. In other words, it is a count of the number of words that appear in both documents. Mathematically, it can also be thought of as the dot product between two binary vectors.

To normalize similarity for differences in the document lengths we introduce the *dice coefficient*. The dice coefficient is the matching coefficient divided by the average number of non-zero entries in both vectors. If the two documents share no words, the dice coefficient will be zero; if they are exactly the same, the dice coefficient will be one.

The *Jacard coefficient* is an alternative metric that also normalizes for document lengths. It is calculated by first calculating the matching coefficient, and then dividing that by the total number of non-zero elements in both documents. In other words it is the number of common words between both documents divided by the total number of words that appear in either.

The *overlap coefficient* is the matching coefficient divided by the number of unique words in the smaller of the two documents. For example if a short document contains 50 unique words, all of which are used in a second longer document containing 200 words, the overlap coefficient will be 1 despite the fact that the second document has many other words that are unique to it. So this metric has a clear bias that may, under some circumstances, be desirable.

Now, consider real-valued vectors. These vectors are more expressive than binary vectors and contain term frequency information. One standard metric we could apply to assess document similarity is the cosine metric. Given two document vectors x and y the cosine between the documents is:

$$\frac{xy^{\mathrm{T}}}{\| x \| \| y \|}$$

where $\| x \|$ is the norm of x. The Euclidean metric is another effective distance metric:

$$\| x - y \|$$

Both of these metrics can be used to compare documents to each other.

3.3 Some words are more important for document similarity

All words are not equally important in assessing document similarity. For example if two words share the word "the" it does not imply a great deal about the similarity between these two documents. On the other hand, it is probably much more significant if two documents share the word "transduction". Methods that identify and emphasize the valuable words, and de-emphasize the less relevant words will improve overall performance.

In general frequent words provide less information about the semantic content of a document than rare words. The frequency of a particular word is typically about proportional to the inverse of its rank in frequency:

$$f \sim \frac{1}{r}$$

Zipf popularized this empirical observation about language and literature (Zipf 1929). So the second most common word will be observed half as frequently as the most common word. The one hundredth most common word will be observed only one hundredth as frequently as the most common word. The implication of this is that there are a few very commonly observed words, but many rarely observed words.

It is the rare words that have the most semantic value however. In Figure 3.2 we have plotted the document frequency of words taken from a corpus of documents containing 15,495 documents pertinent to *Drosophila Melanogaster* (fruit fly) available at FlyBase (*www.flybase.net*) (FlyBase 2003). After excluding words with document frequencies more than ten thousand and fewer than four, we obtain the resulting histogram. Only 20 words appear in more than 5000 documents. Most of these words are very general words such as "at", "which", and "melanogaster". Most words appear in fewer than 25 documents. These are the words that have functional relevance. Some examples include "sycytium", "dyneins", and "ribosylation".

The consequence of these observations is that the rare words are more critical in terms of assessing document similarity. Therefore, schemes that weight words according to the inverse of their frequency may be more effective.

3.4 Building a vocabulary: feature selection

When building up the vocabulary to use for building document vectors, there are some important considerations. One approach is to include counts

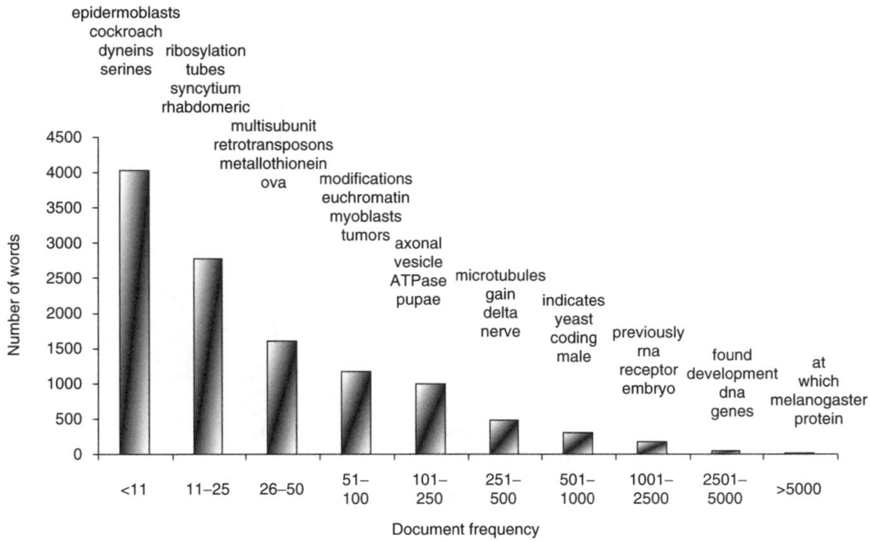

Figure 3.2 *Histogram of words as a function of document frequency in a Drosophila corpus.* In this plot it is apparent that there are just a few words with a very high document frequency, while there are thousands of words with a very low document frequency. The frequently occurring words are not very informative. The examples here are words like "at", "which", and "protein"; these words offer little insight about the content of a document. The rare words such as "ribosylation" and "dyneins" are specific biological concepts whose presence in a document is quite informative.

of all possible words. However, such a vocabulary may be intractably large. In addition, some words, especially the more common ones, may not be helpful in terms of assessing document similarity.

One solution is to use a stop list containing commonly used function words that are valuable in written English, but unhelpful in the setting of document similarity assessment. Such words may include frequently used pronouns, articles, and prepositions, among other words. A short stop list adapted from Manning and Schutze is presented in Table 3.1 (Manning and Schutze 1999).

Another equally effective strategy is to simply remove words that occur either extremely frequently or rarely. A word that appears in many documents is likely to be a function word that does not help in document comparisons; it is probably not semantically specific enough to be helpful. A word that shows up in a few documents may be the consequence of a typographical error, a web site name, or an extremely obscure term that is rarely used. There are often many such words in large corpora that greatly increase the size of the vocabulary and the dimensionality of word vectors without adding significant information. Words such as these may also not be helpful in document comparisons. This simple strategy has been shown to be very effective under many circumstances (Yang and Pedersen 1997).

Table 3.1. *A list of some common stop words.*

a	hers	their
also	his	there
an	how	these
and	I	they
as	if	those
at	in	through
be	it	to
but	its	until
by	me	we
can	my	what
come	of	when
could	on	where
do	one	which
for	or	while
from	our	who
go	say	with
have	she	would
her	that	you
here	the	your

Stemming is yet another strategy that can be employed to reduce the size of the vocabulary. Suffixes are often used to modify words. For example, the word “gene” in its plural form is “genes”. It is sometimes valuable to recognize that these two words are derived from the same root word. Stemming reduces words to their roots. In this way the vocabulary size is reduced as multiple forms of the same word are reduced to a single word. One common approach to stemming is Porter’s algorithm (Porter 1980). Based on a series of ad hoc rules, words are truncated until their root form is achieved.

Another more extreme strategy to reducing the size of the vocabulary is to use only a predefined set of words. For example, one group limited their analysis to words that they were certain pertained only to biological function (Glenisson, Coessens et al. 2004). They restricted their document vectors to include only words and phrases that were predefined in functional vocabularies such as Gene Ontology (Ashburner, Ball et al. 2000). This approach requires the availability of preexisting controlled vocabularies. One of the disadvantages of this approach is that it can be too restrictive, and valuable words that are not explicitly included in the controlled vocabulary may be lost.

3.5 Weighting words

As presented previously, document similarity can be assessed by comparing term frequency vectors. However, term frequency vectors give equal weight to all terms included in the vocabulary. In practice, more effective document similarity measures employ term weighting schemes. These schemes can improve performance by replacing term frequencies with weighted frequencies that emphasize rare terms and dampen frequency. Most of these weighting schemes are practical methods that work well empirically, but are poorly grounded in theory.

The document frequency of a word suggests the importance of a word. As we described above, rare words are likely more informative. So a good weighting scheme should give terms weights that are inversely proportional to the document frequency of the term. A common strategy is to weight terms by a factor $\log_2(N/df_i)$. If a word shows up in every document, then the document frequency is N, and the term receives a weight of zero. On the other hand, if the word appears in a single document only, then the term receives the maximal weight of $\log_2 N$.

Term frequency tells us how valuable a particular term is at describing the content of a document. The more times a word appears in a document, the more likely it is central to the meaning of the document. So any weighting scheme should be proportional to the term frequency. However, while the first occurrence of a term is very indicative about the meaning of a document, each additional occurrence is much less meaningful. Therefore, the most effective weighting schemes often actually dampen the increasing term frequencies. One common scheme is to replace term frequencies with $1 + \log_2(tf_{ij})$ for non-zero term frequencies. So a term appearing once in a document will have a weight of one, but a term appearing 16 times will only have a weight of five. This keeps similarity from becoming skewed by excessive use of individual terms.

So in the weighted word vector the term frequencies, tf_{ij}, are replaced by weights w_{ij} that can be calculated by the following formula:

$$w_{ij} = \begin{cases} [1 + \log(tf_{ij})] \log(N/df_i) & tf_{ij} > 0 \\ 0 & tf_{ij} = 0 \end{cases}$$

So a given document would be represented instead as:

$$d_j = \langle w_{1,j}, w_{2,j}, w_{3,j}, \ldots, w_{N,j} \rangle$$

These weighted word vectors could be compiled into a matrix W, which could be used as an alternative to the word-document matrix A.

There are other equally effective weighting schemes based on the same principles of weighing rare terms and dampening term frequencies. These schemes are generally referred to as term frequency-inverse document frequency (*tf/idf*) weighting.

3.6 Latent semantic indexing

Up until this point we have talked about reducing the number of dimensions in word vectors by removing unimportant words (feature selection and removal of stop words) and have talked about emphasizing certain dimensions by weighting words differentially. Latent semantic indexing (LSI) is a dimension reduction method that works by creating new composite dimensions that are linear combinations of the word dimensions in weighted word vectors (Homayouni, Heinrich et al. 2005). In practice, latent semantic indexing is identical to principal components analysis; this is described as an application to gene expression data in Chapter 2.

Latent semantic indexing is used to transform documents represented in traditional word vector space, where each dimension is a weighted word count, to a space of latent dimensions. Each of the latent dimensions is a normalized linear combination of the word dimensions. In theory the latent dimensions represents semantic meaning instead of just word counts. As in PCA, the latent dimensions are chosen so that they are orthogonal to each other. They are also selected so that they are mutually uncorrelated. Since the latent dimensions are uncorrelated with each other, they aggregate words that are correlated; these are words that frequently co-occur together in documents. The result is that the latent dimensions can actually represent concepts. For example, since the words "cycle", "phase", "cyclins", and "mitosis" are words that might be used together in documents germane to the biology of the cell cycle, these word dimensions might contribute to the same latent dimension (see Figure 3.3).

The advantage of LSI is in converting word vectors to a smaller set of latent dimensions that account for word co-occurrences. For example, two documents may use completely different words and appear dissimilar in word vector space. However, in fact, they may be describing the same concept using different keywords. Latent semantic indexing might define latent dimensions that are linear combinations of words in both documents. Therefore in latent space the documents may actually look similar. In practice latent semantic indexing has been observed to improve information retrieval.

Latent dimensions are created by diagonalizing the covariance matrix. Given a word-document matrix A (or a weighted word document matrix W) where each entry is a weighted word count for the document, the word covariance matrix can be calculated as

$$X = (A - \bar{A})(A - \bar{A})^{\mathrm{T}}$$

where $\bar{A}$ is the average of the weighted counts for each word, and X is the $n \times n$ dimensional word covariance matrix. The covariance matrix, X, can be diagonalized by calculating the eigenvectors v_{i} and eigenvalues d_{i}:

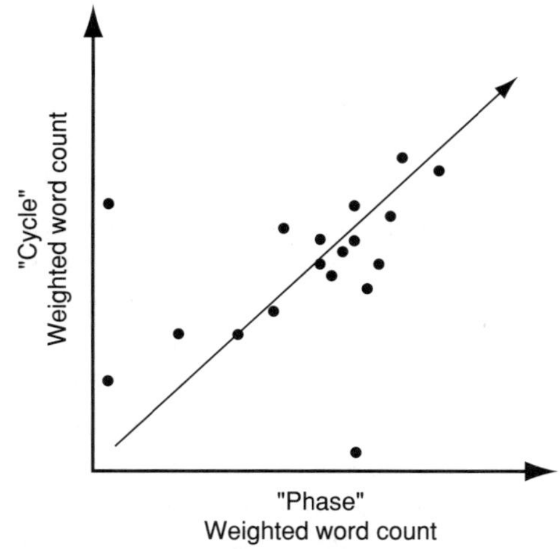

Figure 3.3 *Latent semantic indexing.* Here we have plotted documents (black dots) as a function of their weighted word counts of the words "cycle" and "phase". Few documents may use these words in isolation (black dots along the axes). Documents about cell cycle biology, however, would frequently use both. These words have strong covariance. An example of a latent dimension is given with the diagonal arrow; it captures the greatest variance in the data. Documents that are relevant to cell cycle physiology will have large values in this direction.

$$\begin{pmatrix} d_1 & & & 0 \\ 0 & & & \\ & \cdots & & \\ & & d_{n-1} & 0 \\ 0 & & 0 & d_n \end{pmatrix} \begin{pmatrix} v_{1,\,1} & & v_{n-1,\,1} & v_{n,\,1} \\ v_{1,\,2} & & v_{n-1,\,2} & v_{n,\,2} \\ & \cdots & & \\ & & & \\ v_{n,\,1} & & v_{n,\,n-1} & v_{n,\,n} \end{pmatrix}$$

$$= X \begin{pmatrix} v_{1,\,1} & & v_{n-1,\,1} & v_{n,\,1} \\ v_{1,\,2} & & v_{n-1,\,2} & v_{n,\,2} \\ & \cdots & & \\ & & & \\ v_{n,\,1} & & v_{n,\,n-1} & v_{n,\,n} \end{pmatrix}$$

where each of the columns in the matrix V are eigenvectors; they are orthogonal to each other and are normalized vectors. Each vector represents a new latent dimension. Diagonalizing the matrix is tantamount to rotating the coordinate space with the matrix V, so that the covariance terms are zero. The new covariance matrix is D, where the off-diagonal terms are zero. The terms in the diagonal matrix D are the eigenvalues; they represent the variance of the corresponding dimension. The dimensions associated with greatest variance are considered most informative. In practical applications of LSI the eigenvectors with the largest associated variance are used, and other eigenvectors are eliminated. The result is a dimensional reduction. Documents or gene-text profiles can be transformed into the latent space by multiplying them with the matrix of the top eigenvectors:

$$A_{\mathrm{LSI}} = V_m^{\mathrm{T}} A$$

Here, A_{LSI} is the vector of documents in m-dimensional LSI space, and V_m is a matrix consisting only of the top m eigenvectors in the matrix V corresponding to the greatest variance. Document or gene similarity can be quantified with the cosine of the angle between vectors in the reduced latent space.

To illustrate LSI, we obtained a list of 15,495 articles relevant to *Drosophila* genes from FlyBase. We looked only at the words with document frequencies of at least 100 and less than 5000. In the corpus of 15,495 articles; there were 2021 such words. After constructing weighted word vectors, we calculated a 2021 × 2021 covariance matrix, and diagonalized it. We then selected the top 100 eigenvectors to create a latent dimensional space. A total of 26.4% of the total variance is captured in these 100 dimensions (Figure 3.4). About 3/4 of the variance or information is completely discarded in this representation. This represents a twenty-fold reduction in dimensionality of the data. In Section 3.10 we will show that despite this marked reduction in dimensionality, there can actually be improvements in query performance.

3.7 Defining textual profiles for genes

Until this section we have been discussing document vectors. In this section we will talk about converting document vectors into gene vectors.

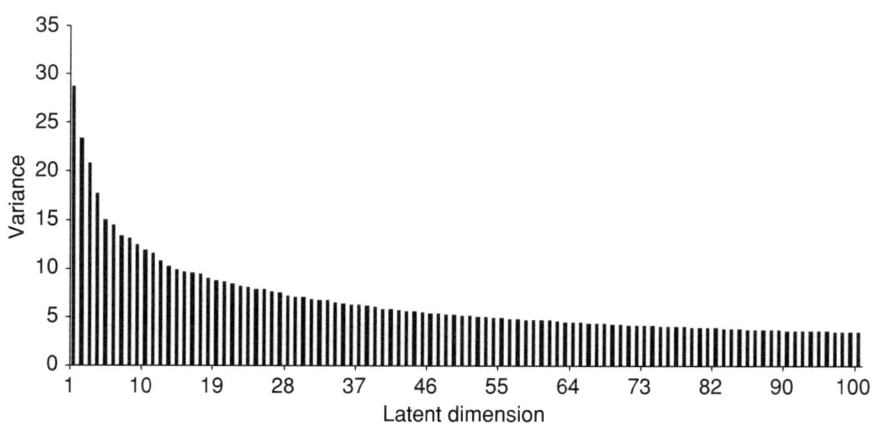

Figure 3.4 *Variance as a function of latent dimension.* In this example there are a total of 2021 latent dimensions. The first 100 capture about 1/4 of the total variance. Variance (or eigenvalue) per latent dimension is a measure of the total information contained in that dimension. Here we have plotted total variance for each dimension in descending order; notice that the information per dimension drops off quickly.

As a starting point, we build reference indices connecting genes to document. Manually curated indices are available from online genomic resources such as FlyBase (FlyBase 2003). Alternatively such an index can be built up by tabulating articles that simply mention a gene. These and other indices were discussed in detail in Chapter 1. We can define a matrix G that contains the document gene index. For each gene we define a vector in the space of documents; reference documents for the gene are set to one, while other documents are set to zero. We assemble the vectors into a matrix, G, where each column corresponds to a gene, and each row corresponds to a specific document in the corpus. The value of each entry G_{ij} is one only if document i refers to gene j. Shatkay and colleagues took this approach to define and cluster genes (Shatkay, Edwards et al. 2000).

In the remainder of this chapter we use a reference index from *FlyBase* that contains 15,495 documents to create gene profiles. These articles were connected to a set of 14,732 genes. The median number of references per gene is 1, while the mean number of references per gene is 6.6. This difference between the mean and median number of references per gene is a consequence of the skewed distribution of literature that was discussed in detail in Chapter 1.

Given a collection of document references for a specific gene, we can count up all the constituent words and define a vector for that gene. This is the textual profile for the gene. In this case, we have a matrix H where each column corresponds to a specific gene, and each row corresponds to a vocabulary word. Each entry, H_{ij}, corresponds to the number of times a word i appears in documents pertinent to gene j. In general, the matrix of textual profiles H can be derived by multiplying the word-document matrix A and the reference index G:

$$H = A \times G$$

One research group took this approach when they organized human genes into functional categories automatically based on the scientific literature about these genes (Chaussabel and Sher 2002). After creating word vectors for each human gene, they used hierarchical clustering to organize those genes into functional categories. Another group took a similar approach with yeast and human genes and implemented an online resource for text profiling of genes (Glenisson, Coessens et al. 2004).

The textual profiles that we use in this text are slightly different from the above. We use the average of the weighted word vectors for document references to a gene to create the textual profile. In matrix form:

$$H = \mathrm{W} \times \overline{\mathrm{G}}$$

where W is the weighted word document matrix and $\overline{G}$ is the normalized reference index, where values in each of the columns are reduced so that they sum to one. The result is a textual profile for each gene in which words are given differential weights depending on their relevance to the gene. This is a very simple, but effective, textual representation for each gene.

3.8 Using text like genomics data

The scientific text, after being converted into matrix form, can be used as another type of biologically meaningful genomic data. The textual profile of a gene gives us information about the gene much the same way that gene expression or sequence data does. We can use pairwise word vector similarity to assess gene similarity as we can use pairwise sequence alignment or correlation between gene expression profiles. Alternatively we can look at keywords or phrases that are associated with a gene the same way we can look at the presence of sequence motifs or increased expression under specific conditions to understand the gene's function.

To demonstrate that the scientific literature is related to biological function we focus on the *Drosophila melanogaster* literature and the well studied gene *breathless*. Consider the *FlyBase* reference index described in Sections 3.6 and 3.7. In this section, we exclude only those words with document frequencies more than 10,000 and fewer than four. Then the documents are converted into weighted vectors of words. We average each of the gene's referring document vectors together into a textual profile for each gene. These gene vectors are equivalents to gene vectors in matrix H.

The *breathless* gene has 103 references in this index. It is a fibroblast growth factor (FGF) receptor that is involved in a signal transduction pathway and the control and differentiation of tracheal cells (Dossenbach, Rock et al. 2001). In Table 3.2 we list the terms with greatest weight in the gene's word vector. Notice that these keywords include the name of the gene and many other words that correspond to the gene's function. Words like "tracheal" and "migration" indicate the physiological function of the gene, while other words like "signaling", "FGF", and "receptor" suggest the biochemical functions of the gene.

We can calculate how similar the word vector of *breathless* is to other gene word vectors by calculating the cosine between those vectors. We look only at the genes with a sufficient number of references to insure that our results are reliable, and not the function of sparse data. The *breathless* word vector has a mean cosine similarity score of 0.23 with the other 2325 genes with five or more document references; it has a median cosine similarity of 0.20. More than 80% of genes have word vectors with cosine similarities

Table 3.2 *Keywords for the two similar genes in Drosophila: breathless (103 abstracts) and heartless* (75 abstracts).

Breathless keywords	tf/idf score	*Heartless* keywords	tf/idf score
tracheal	6.9	FGF	4.6
FGF	5.5	mesoderm	3.6
migration	4.4	receptor	3.4
receptor	3.2	heartless	3.0
breathless	3.0	signaling	3.0
branching	2.7	fibroblast	2.8
branches	2.6	migration	2.7
signaling	2.6	mesodermal	2.7
cell	2.5	muscle	2.7
cells	2.4	cells	2.4
tyrosine	2.0	growth	2.0
fibroblast	1.9	cell	1.9
sprouty	1.7	sprouty	1.8
growth	1.6	factor	1.8
branch	1.5	twist	1.7

less than 0.3, while 98% of genes have cosine similarities less than 0.5. This distribution is illustrated in Figure 3.5.

The *breathless* gene has a homolog called *heartless*; it has a similar function as *breathless*, but controls the migration and differentiation of mesodermal cells instead of for tracheal cells. *Heartless* has 75 document references. Since these genes are so similar, we would certainly expect that they would have similar word vectors as well. In Table 3.2 we have also listed the highest scoring keywords for *heartless*. Notice that many of the keywords are very similar or identical to *breathless* keywords. Instead of "tracheal", however, we have the more appropriate physiological term "mesodermal". The cosine similarity score for the two word vectors of these genes is 0.73. This is a much higher score than the score between *breathless* and almost all the other genes.

Another point to consider is whether there is a correlation between textual data and other biological data. For example, we can consider the amino acid sequences of gene protein products. Using BLAST we compared the amino acid sequence of the *breathless* protein product to all of the other known gene protein products in *Drosophila*. We obtained 121 statistically significant hits. These protein sequences corresponded to 67 unique *Drosophila* genes. Of these, we found 32 genes with five or more references. The mean word vector cosine similarity between *breathless* and the genes with similar sequences is 0.33; the median vector cosine similarity is also 0.33. These similarity scores are significantly greater than the average similarity score of all genes (see Figure 3.5). There is a relationship between

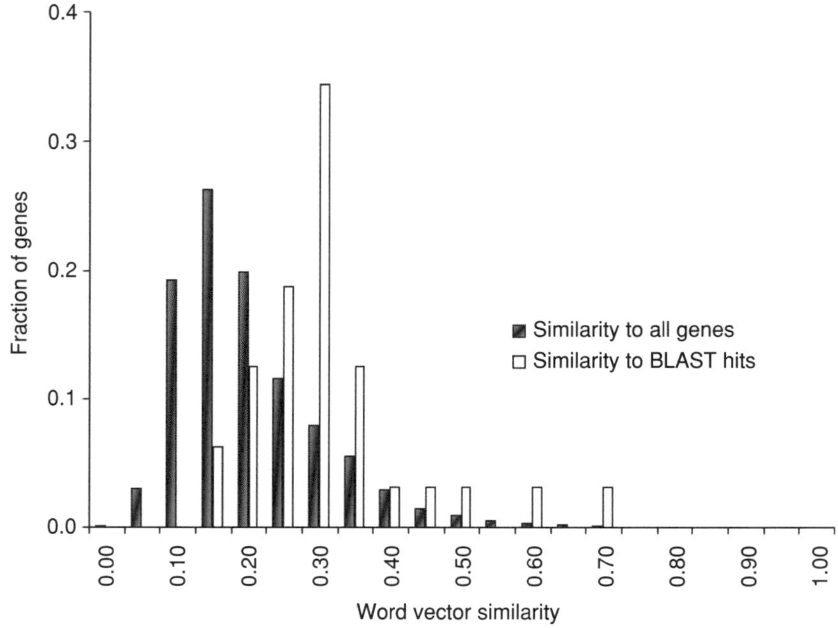

Figure 3.5 *Word vector similarity between other Drosophila genes and the breathless gene.* Here we plot a histogram of genes as a function of word vector similarity. In addition we have plotted a separate histogram of word vector similarity between the *breathless* gene and other *Drosophila* sequences that are similar to it identified with BLAST.

sequence similarity scores and the word vector similarity as depicted in Figure 3.6. The more similar the hit sequence is, the more likely it shares function with *breathless* and has similar textual content. The correlation between BLAST scores and word vector similarity is 0.68. The best example is the *heartless* gene with dramatic sequence and word vector similarity to *breathless*. After removing this potential outlier, the correlation is still 0.42.

We notice a similar relationship when we look at gene expression data. Here we look at data from a comprehensive expression assay that followed genes throughout the life cycle of *drosophila* (Arbeitman, Furlong et al. 2002). This experiment consists of 85 measurements at different time points throughout fruitfly development of 3987 unique genes. Expression values for genes were recorded as log ratios between expression at each value and a baseline. Gene expression profiles can be compared to each other through cosine similarity. We focus on a subset of genes with 25 or more document references and greater than an expression variance of 0.3 across the 85 conditions. This insures that the genes we are looking at have significant changes in expression and sufficient text available. There are 212 genes that fit these criteria. In Figure 3.7 we plot word vector similarity to

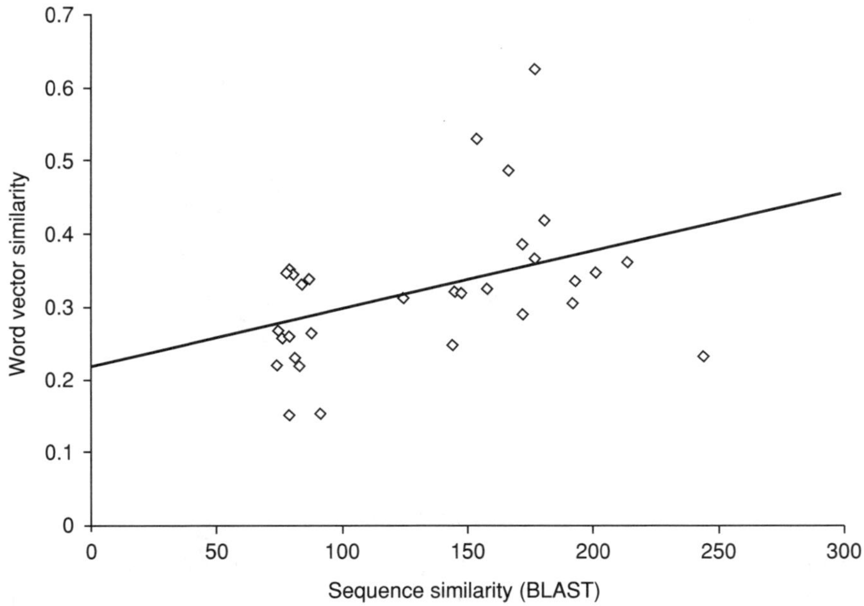

Figure 3.6 *Word vector similarity to breathless gene versus sequence similarity.* We have plotted for each gene its word vector cosine similarity to the *breathless* gene as a function of its BLAST protien sequence similarity score to the *breathless* protein sequence. Only BLAST hits to *breathless* are plotted. Here we have removed the *heartless* gene, which has extreme similarity in sequence and text.

breathless as a function of gene expression similarity for these genes. There is a correlation of 0.35 between expression and text similarity.

These are very coarse examples; but they demonstrate that there is a relationship between textual similarity and biological similarity. These experiments suggest that textual analysis might be helpful in the analysis of biological data. In Chapter 4 we will present methods to exploit textual similarity to enhance sequence analysis. In Chapters 5 and 7 we will present methods to exploit textual similarity to enhance expression analysis.

The representation of text that we introduce in this chapter is very simple; all of the documents for a single gene were averaged together and no higher level processing was attempted. Yet, even in such an unsophisticated approach, biological similarity is, to some extent, preserved. The levels of correlation in these examples are low; they are comparable to the correlations that we would obtain by comparing any two very different forms of genomics data. Investigators have conducted more detailed evaluations of the performance of representations like textual profiles and have compared different implementations and weighting schemes (Glenisson, Coessens et al. 2004).

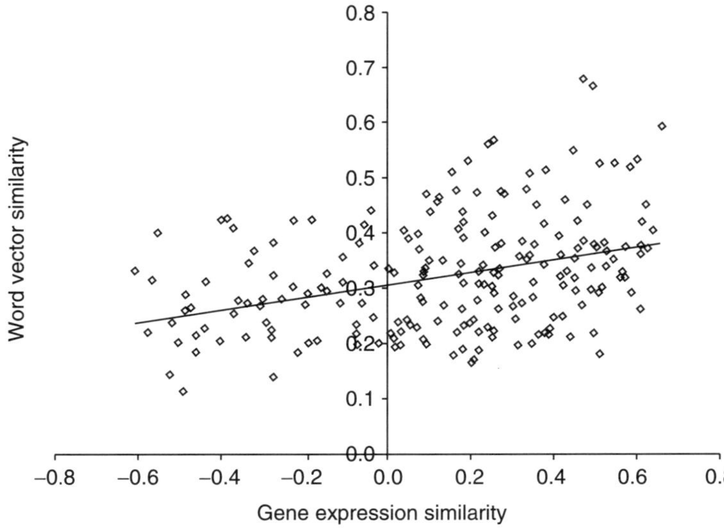

Figure 3.7 *Word vector similarity to breathless gene versus gene expression similarity.* We have plotted for each gene its word vector similarity to the *breathless* gene as a function of its gene expression similarity to *breathless*.

3.9 A simple strategy to assigning keywords to groups of genes

One very useful application of the word vector representation for genes, besides its use as another set of genomic data, is its ability to be used to assign keywords. In the previous section we demonstrated that the word vector for a gene, created by averaging weighted word vectors of articles referring to the gene, provide an avenue for selecting descriptive keywords. We list the keywords for the *breathless* gene and the *heartless* gene in Table 3.2. The highest valued words in this vector have high values because they are (1) frequent and appear in many of the referring documents and (2) they are rare in the corpus and therefore heavily weighted words. These words are ideal keywords and provide meaningful clues about the function of a gene.

This concept can be extended to groups of genes as well. Given a group of genes that are produced by an experimental assay (such as a gene expression cluster), the word vectors for each gene can be averaged together. The words associated with the functions that are recurring themes in that group will receive high values, and will be effective keywords. There are other more sophisticated strategies for keyword assignment as well, which we will explore further in future chapters.

3.10 Querying genes for biological function

Often times we want to search documents, or genes, for certain concepts using keywords. One very important type of query in biology is searching for genes that have a particular biological function. One of the great efforts of the genomics era is identifying and cataloging genes with specific functions. We have already demonstrated how functionally valuable keywords can be obtained for a gene using textual profiles. In this section we discuss how we can query genes for specific functions with keywords.

We can use the word vector formulation that we have been using for genes and documents to facilitate such queries. A keyword query can be formulated as a pseudo-document vector where each word in the query is assigned one in the word vector, all other vector entries are zero. One queries a database of documents by calculating the cosine between that vector and all of the weighted word vector representations of the documents in the database. The highest scoring documents are returned in the query. This same strategy can be applied to query gene textual profiles.

To demonstrate with an example, we query *drosophila* genes for the functions "signal transduction" or "eye morphogenesis". Ideally these queries should be able to find genes with the appropriate biological function. We tested this possibility in standard word vector space and subsequently in latent space.

We used the FlyBase corpus and reference index described in Sections 3.6 and 3.7. We used those words with document frequencies of at least 100 and less than 5000 to identify 3021 vocubulary words. We average weighted word vector documents that are references to a gene to create a textual profiles for that gene. We discarded all of the genes with only a single reference. We divide up the genes into well-documented genes (more than ten references) and poorly documented genes (ten or fewer references). A total of 4609 genes have ten or fewer article references; a total of 1276 genes have more than ten references. Since we were testing queries for signal transduction genes and eye morphogenesis genes, we assembled a gold standard for each query. Gene Ontology annotated 260 genes as relevant to signal transduction. Of these, 93 have greater than ten references; 127 have more than a single reference but ten or fewer references. Gene Ontology annotated 43 genes as relevant to eye morphogenesis. Of these, 26 have greater than ten references; 13 have more than a single reference but ten or fewer references. Effective queries should be able to separate these genes from the rest. We formulated pseudo-vectors for both queries where all of the words except the ones in the query quotes are zero valued. The cosine between query pseudo-vectors and gene textual profiles were calculated. Then query sensitivity and specificity values were calculated for different cosine cutoffs. A stringent cutoff would result in low sensitivity but high specificity.

In Figure 3.8 we display query performance separately for the well referenced and poorly referenced genes. For the "signal transduction"

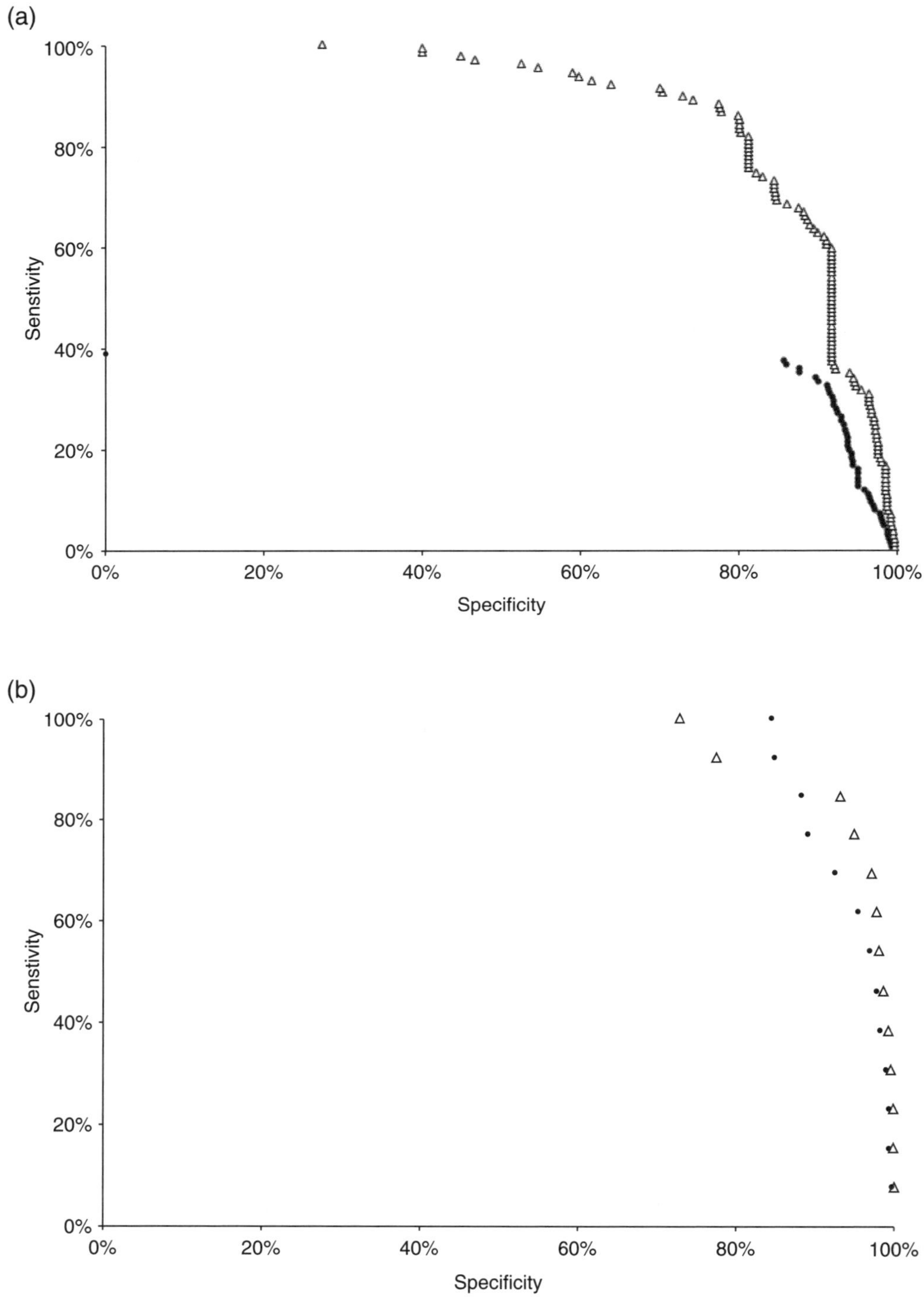

Figure 3.8 *Keyword queries in word vector space versus LSI space.* The plots above are sensitivity–specificity plots for queries formulated in word vector space (dark dots) and LSI space with 100 dimensions (open triangles). An ideal query achieves 100% sensitivity at 100% specificity. (a) Querying genes with fewer references for "signal transduction". (b) Querying genes with fewer references for "eye morphogenesis".

(c)

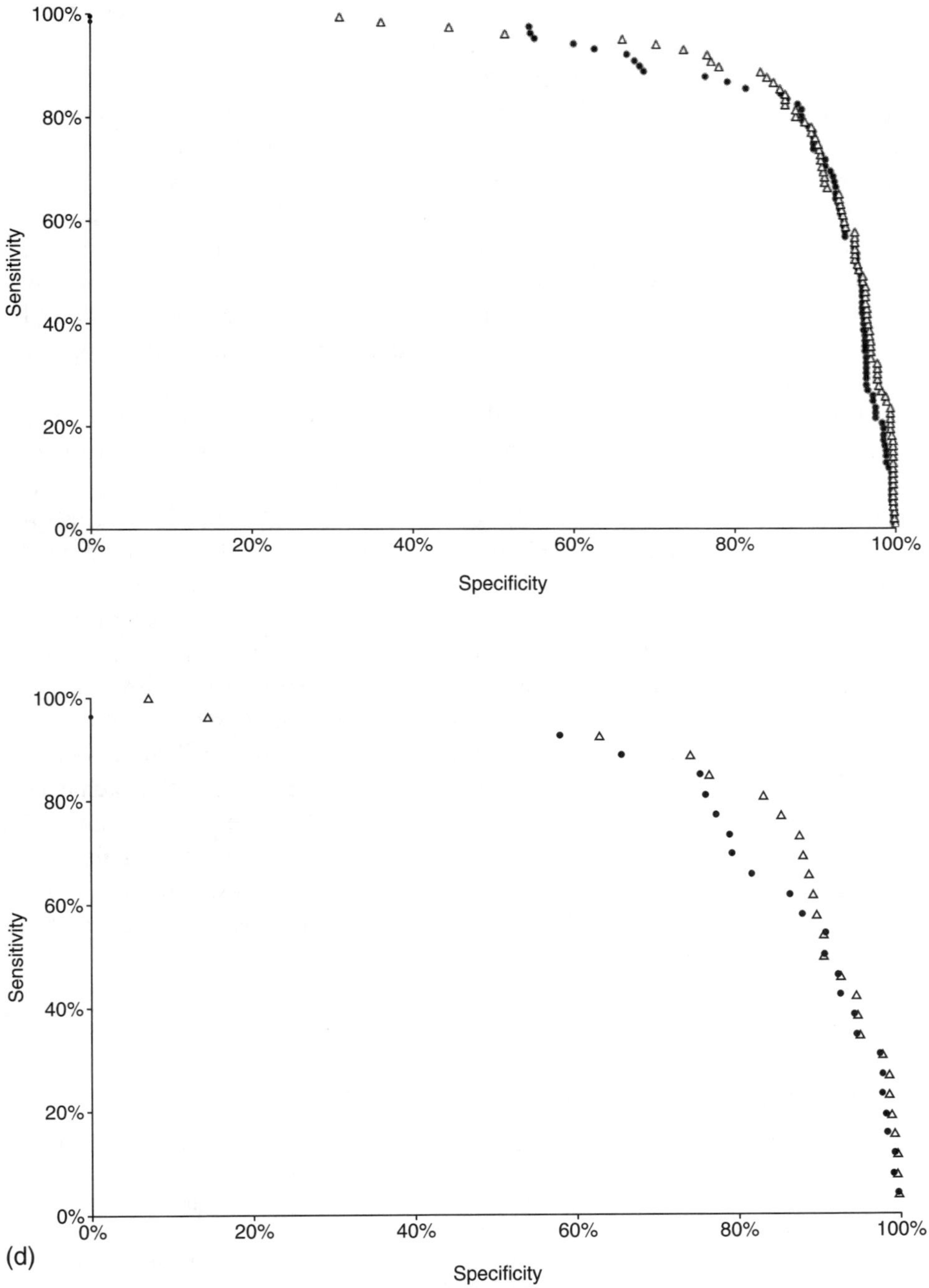

(d)

Figure 3.8 (c) Querying genes with more references for "signal transduction". (d) Querying genes with more references for "eye morphogenesis".

query this strategy achieves 38% sensitivity at 95% specificity for all genes studied; it achieves 52% sensitivity at 90% specificity. For the "eye morphogenisis", the query achieves 54% sensitivity at 95% specificity overall; it obtains 67% sensitivity at 90% sensitivity.

We noted previously that latent semantic indexing often improves information retrieval queries. To demonstrate, we implemented these queries in latent space as well. Standard weighted word vectors (2021 dimensions) were converted into latent word vectors (100 dimensions) by multiplying with matrix V_m. This conversion is detailed at the end of Section 3.6. Latent document vectors were averaged to create latent textual profiles for all genes. The query vectors were similarly transformed into latent space by multiplying by matrix V_m as well. The cosine of the latent query vector and the latent document vectors can be calculated, and the highest scoring genes are returned in a query.

In both cases, latent semantic indexing achieves better query performance than standard word vectors. For the "signal transduction" query, this strategy improves sensitivity from 38% to 52% at 95% specificity; it improves sensitivity from 52% to 73% sensitivity at 90% specificity. For the "eye morphogenesis" query, this strategy achieves 54% sensitivity at 95% specificity; it improves sensitivity from 67% to 82% at 90% specificity. The performance improvements from latent semantic indexing are most notable when there is a paucity of literature (Figure 3.8). However in cases where there are many references, the performance between the two methods is more equivocal.

Incidentally, the sensitivities and specificities in both cases are not nearly high enough for accurate gene annotation. If, for example, 100 genes out of 1000 have a specific biological function, then even a query that achieves 90% sensitivity at 90% specificity would return 90 true positives and 90 false positives. So a full half of the presumed positives are errors. This level of accuracy is simply too low. Accurate gene annotation will require more sophisticated methods to achieve greater performance. We will explore this question in Chapter 8.

References

Altman, R. B. and S. Raychaudhuri (2001). "Whole-genome expression analysis: challenges beyond clustering." *Curr. Opin. Struct. Biol.* **11**(3): 340–7.

Arbeitman, M. N., E. E. Furlong, et al. (2002). "Gene expression during the life cycle of Drosophila melanogaster." *Science* **297**(5590): 2270–5.

Ashburner, M., C. A. Ball, et al. (2000). "Gene ontology: tool for the unification of biology. The Gene Ontology Consortium." *Nat. Genet.* **25**(1): 25–9.

Chaussabel, D. and A. Sher (2002). "Mining microarray expression data by literature profiling." *Genome. Biol.* **3**(10): RESEARCH0055.

Dossenbach, C., S. Rock, et al. (2001). "Specificity of FGF signaling in cell migration in Drosophila." *Development* **128**(22): 4563–72.

Eisen, M. B., P. T. Spellman, et al. (1998). "Cluster analysis and display of genome-wide expression patterns." *Proc. Natl. Acad. Sci. USA*. **95**(25): 14863–8.

FlyBase (2003). "The FlyBase database of the Drosophila genome projects and community literature." *Nucleic Acids Res.* **31**(1): 172–5.

Glenisson, P., B. Coessens, et al. (2004). "TXTGate: profiling gene groups with text-based information." *Genome. Biol.* **5**(6): R43.

Hersh, W. (2003). *Information Retrieval: A Health and Biomedical Perspective*. New York, Springer-Verlag.

Homayouni, R., K. Heinrich, et al. (2005). "Gene clustering by Latent Semantic Indexing of MEDLINE abstracts." *Bioinformatics* **21**(1): 104–15.

Manning, C. M. and H. Schutze (1999). *Foundations of Statistical Natural Language Processing*. Cambridge, The MIT Press.

Porter, M. F. (1980). "An algorithm for suffix stripping." *Program* **14**: 130–7.

Rosenfeld, R. (2000). "Two decades of statistical language modeling: where do we go from here?" *Proceedimgs of the IEEE* **88**(8): 1270–1278.

Shatkay, H., S. Edwards, et al. (2000). "Genes, themes and microarrays: using information retrieval for large-scale gene analysis." *Proc. Int. Conf. Intell. Syst. Mol. Biol.* 8(10): 317–28.

Yang, Y. and J. P. Pedersen (1997). "A Comparative Study on Feature Selection in Text Categorization. Proceedings of the Fourteenth International Conference on Machine Learning." *International Conference on Machine Learning*. Morgan Kaufmann, San Francisco pp. 412–20.

Zipf, G. K. (1929). "Relative frequency as a determinant of phonetic change." *Harvard Studies in Classical Philology* **40**: 1–95.

4 Using text in sequence analysis

Text about genes can be effectively leveraged to enhance sequence analysis (MacCallum, Kelley et al. 2000; Chang, Raychaudhuri et al. 2001; McCallum and Ganesh 2003; Eskin and Agichtein 2004; Tu, Tang et al. 2004). Most of the emerging methods utilize textual representations similar to the one we introduced in the previous chapter. To analyze sequences, a numeric vector that contains information about the counts of different words in references about that sequence can be used in conjunction with the actual sequence information.

Experienced biologists understand the value of using the information in scientific text during sequence searches, and commonly use scientific text and annotations to guide their intuition. For example, after a quick BLAST search, a trained expert might quickly look over the hits and their associated annotations and literature references and assess the validity of the hits. The apparently valid sequence hits can then be used to draw conclusions about the query sequence by transferring information from the hits.

In most cases, the text serves as a proxy for structured functional information. High quality functional annotations that succinctly and thoroughly describe the function of a protein are often unavailable. Defining appropriate keywords for a protein requires a considerable amount of effort and expertise, and in most cases, the results are incomplete as there is an ever-growing collection of knowledge about proteins. So, one option is to use text to compare the biological function of different sequences instead.

There are different ways in which the functional information in text could be used in the context of sequence analysis. One possibility is to first run a sequence analysis algorithm, and then to use text profiles to summarize or organize results. Functional keywords can be assigned to the whole group of hit sequences. Additionally, given a series of sequences, they can be grouped according to like function. In either case, quick assessment of the content of text associated with sequences offers insight about exactly what we are seeing. These approaches are particularly useful if we are querying a large database of sequences with a novel sequence that we have very little information about. In these cases text analysis is used only to describe results obtained from a sequence analysis query.

On the other hand text analysis could be used in addition to sequence analysis to actually help identify homologous genes. That is sequence-based database searches could actually leverage the text as well. Remotely homologous genes are genes that have diverged in evolution early and consequently sequence similarity is only modest, but often these genes continue to have the same protein structure and function. As a result of the sequence divergence, traditional sequence similarity measures perform poorly. These sequences are said to be in the "twilight zone". One solution to recognizing remotely homologous genes is to use the functional information in the literature to help recognize whether a sequence with questionable similarity may be homologous. Iterative sequence analysis algorithms can be modified so that at each iteration of a sequence search, analysis of the textual profiles are used to influence the intermediate results before the next iteration commences.

Text analysis can also be used in functional classification algorithm that uses both sequence and text simultaneously to assign protein function. Algorithms can look for features of protein sequences that identify its function. When attempting to ascertain if a particular protein has a given function, the textual features of literature references can be used in addition to help identify the function.

These approaches are a few examples of how text can be used with functional genomics data in different ways and to different extents. In later chapters we address using scientific literature to help analyze gene expression data; many of the same principles apply. Whatever strategy is employed, there are a few pitfalls that must be carefully regarded.

One of the great challenges to including literature in analyzing genomics data is that most genes or sequences lack literature altogether. Many have very limited amounts of literature. We must be cautious of weighting text-mining approaches heavily when analyzing these sequences with lack of literature. On the other hand, there are sequences with large amounts of literature. Should one of these sequences be included in the collection of sequences being analyzed, we must be careful not to let its very large body of literature overwhelm the analysis and prejudice the results.

The concepts discussed in this chapter are listed in the frame box. We first introduce the SWISS-PROT database record; besides containing a protein sequence, the record contains valuable text describing the protein and is the most common literature resource used in sequence analysis. Then we discuss strategies to assign literature references to poorly referenced genes by transferring references from genes with similar sequences. We present strategies to assign functional keywords to a gene from sequence similarity hits. We then describe how sequence hits can be organized using their scientific literature. We describe two methods that use textual analysis to help

1) The SWISS-PROT record 2) Transferring gene references to uncharacterized genes based on sequence similarity 3) Summarizing and organizing BLAST hits	a) Assigning functional keywords to sequences b) Clustering hit sequences 4) Recognizing homologous genes 5) Predicting gene function from sequence and text

recognize homology among remotely similar genes. Finally we present some of the recent work in predicting function with both text and sequence information

4.1 SWISS-PROT records as a textual resource

A very valuable textual resource in sequence analysis is SWISS-PROT (Boeckmann, Bairoch et al. 2003). In Figure 4.1 we present selected fields from a typical SWISS-PROT record; this is the amino acid sequence record for *breathless*, the same gene we used as an example in chapter 3.

In this record there are important identifying fields, such as the accession number (AC). In addition, two other extremely valuable text fields are the keywords field (KW) and the comments field (CC). These two fields contain highly distilled textual information about the protein. In this example, we can see that there is information about the protein's physiological function, cellular location, role in development, and biochemistry. Indeed the readily available text of the SWISS-PROT record alone can be used in enhancing sequence analysis.

When protein sequences are submitted to SWISS-PROT, authors include literature references so that users may better understand the sequence and the context in which it was obtained. The reference titles (RT) are listed in the record and the abstract text can be obtained from PubMed using the MedLine identification numbers included with each reference (RX). While many of these references are helpful, they are often a limited subset of the full breadth of available literature and provide only a limited view of the relevant biology of the sequence. For example, there are only four references provided in this sequence record, whereas FlyBase provided some 103 references for the *breathless* gene. In general there is a median of only one article reference per sequence in SWISS-PROT.

Ideally textual profiles for a sequence could be constructed by recognizing the gene that the sequence in the record pertains to, obtaining all of the

```
ID   FGR2_DROME     STANDARD;      PRT;  1052 AA.
AC   Q09147;
DE   (Breathless protein) (dFGF-R1).
GN   BTL OR FR2 OR DTK2.
OS Drosophila melanogaster (Fruit fly) .
RP   SEQUENCE FROM N.A.
RC   STRAIN=Canton-S;
RX   MEDLINE=93321617; ;
RA Shishido E. ,Higashijima S.-I. ,Emori Y. ,Saigo K. ;
RT   "Two FGF-receptor homologues of Drosophila: one is expressed in
RT   mesodermal primordium in early embryos.";
RL Development 117:751-761(1993). RN   [2]
RP   SEQUENCE OF 1-240 FROM N.A.
RX   MEDLINE=92387542; ;
RA Klaembt C. ,Glazer L. ,Shilo B.-Z. ;
RT   "Breathless, a Drosophila FGF receptor homolog, is essential for
RT   migration of tracheal and specific midline glial cells.";
RL Genes Dev. 6:1668-1678(1992). RN   [3]
RP   SEQUENCE OF 267-1052 FROM N.A.
RC   TISSUE=Embryo;
RX   MEDLINE=91184623; ;
RA Glazer L. ,Shilo B.-Z. ;
RT   "The Drosophila FGF-R homolog is expressed in the embryonic tracheal
RT   system and appears to be required for directed tracheal cell
RT   extension.";
RL Genes Dev. 5:697-705(1991). RN   [4]
RP   SEQUENCE OF 868-923 FROM N.A.
RX   MEDLINE=92008631; ;
RA Shishido E. ,Emori Y. ,Saigo K. ;
RT   "Identification of seven novel protein-tyrosine kinase genes of
RT   Drosophila by the polymerase chain reaction.";
RL   FEBS Lett. 289:235-238(1991).
CC   -!- FUNCTION: May be required for patterning of muscle precursor
CC       cells. Would thus appear essential for generation of mesodermal
CC       and endodermal layers, invaginations of various types of cells,
CC       and CNS formation.
CC   -!- CATALYTIC ACTIVITY: ATP + a protein tyrosine = ADP + protein
CC       tyrosine phosphate.
CC   -!- SUBCELLULAR LOCATION: Type I membrane protein.
CC   -!- TISSUE SPECIFICITY: Mesoderm.
CC   -!- DEVELOPMENTAL STAGE: Embryogenesis. DFR2 expression occurs in
CC       endodermal precursor cells, CNS midline cells and certain
CC       ectodermal cells such as those of trachea and salivary duct.
CC   -!- SIMILARITY: Belongs to the fibroblast growth factor receptor
CC       family.
CC   -!- SIMILARITY: Contains 5 immunoglobulin-like C2-type domains.
KW Receptor ;Glycoprotein ;Tyrosine-protein kinase ;ATP-binding ;
KW Transferase ;Phosphorylation ;Transmembrane ;Immunoglobulin domain ;
KW Repeat ;Signal .
SQ   SEQUENCE   1052 AA;  117824 MW;  1E8B980E16DC8D15 CRC64;
     MAKVPITLVM IIAIVSAAAD LGCDYGHHRC YIDVTVENSP RQRHLLSDMD ITLQCVRPMA
     KWFYEDKFQL RATLLRLERA QSGNSGNYGC LDSQNRWYNI SLVIGHKEPV GNDIASFVKL
     EDAPAIPESD LFFQPLNESR SLKLLQPLPK TVQRTAGGLF QLNCSPMDPD AKGVNISWLH
     IDTQILGGRG RIKLKRWSLT VGQLQPEDAG SYHCELCVEQ DCQRSNPTQL EVISRKHTVP
     MLKPGYPRNT SIALGDNVSI ECLLEDSALE PKITWLHKGN ADNIDDLLQR LREQSQLPVD
     VTRLITRMDE PQVLRLGNVL MEDGGWYICI AENQVGRTVA ASYVDLYSPS DTTTVRTTTT
     TTVASPIPTA STGEDNDDDV ENPAADASGG VGPPVFRKEL KRLQHSLSGN TVNLACPVYG
     KANITWTKDK KPLNRELGVY VQKNWTLRFV EATSEDSGLY NCKVCNAWGC IQFDFSVQIN
     DRTRSAPIIV VPQNQTVKVN GSLVMKCTVY SDLHPTVSWK RVVLKNASLD GLQSVEIQNL
     NFTVTNDSVV LTLRNVTFDQ EGWYSCLASS GLGRSNSSVY LRVVSPLPPL EIYALLHAHP
     LGFTLAAITI VALFLLGSAF ITFMLRRLRR EKLLKLRIET VHQWTKKVII YRPGGEEGSG
     CSSGDLQMPV IRIEKQRTTV STTGTGGTDP AQGFNEYEFP LDSNWEIPRQ QLSLGSILGE
     GAFGRVVMAE AEGLPRSPQL AETIVAVKMV KEEHTDTDMA SLVREMEVMK MIGKHINIIN
     LLGCCSQGGP LWVIVEYAPH GNLKDFLKQN RPGAPQRRSD SDGYLDDKPL ISTQHLGEKE
     LTKFPFQIAR GMEYLASRRC IHRDLAARNV LVSDGYVMKI ADFGLARDIQ DTEYYRKNTN
     GRLPIKWMAP ESLQEKKYDS QSDVWSYGVL LWEIMTYGDQ PYPHILSAEE LYSYLITGQR
     MEKPAKCSLN IYVVMRQCWH FESCARPTFA ELVESFDGIL QQASSNPNDA YLDLSMPMLE
     TPPSSGDEDD GSDTETFRET SPLRYQYTYK FN
//
```

Figure 4.1 *SWISS-PROT record for breathless protein sequence.*

known references for that gene, and then building a textual profile from those references, the references in the SWISS-PROT record, and also key fields in the record itself. In practice, however, it is often simpler to construct textual profiles for sequences with only the SWISS-PROT record and the references contained within it.

4.2 Using sequence similarity to extend literature references

One of the basic goals of sequence analysis and sequence similarity searches is to take unknown sequences and to learn something new about them. This is especially the case for newly discovered sequences for which there is a paucity of information. We find similar sequences and we might assume that the known functions of those sequences are potentially relevant to the newly discovered gene or protein.

In many cases genes and proteins are poorly studied and they have no documented references. Sequence similarity can be used to assign new pseudo-references to such proteins. These new references can be used to help understand gene function. They can also be used to facilitate text-based analysis with the sorts of algorithms we introduce in this book. For example, effective analysis of a gene expression data set that has many unstudied genes might be facilitated if the unstudied genes could have putative references assigned.

One possible strategy is to conduct a sequence similarity search with BLAST or some other algorithm against a large database of sequences. Then we can transfer the references from the most similar sequences. Such references might provide valuable insight about the sequence. In an ideal situation a stringent search will find a well-studied homologous sequence with very similar biological function in other organisms, with applicable available references. If such homologous sequences are unavailable, a less stringent search may reveal other similar sequences with protein domains that have vaguely similar molecular function, and references from those sequences will still provide hints of function. Using a very stringent sequence similarity cutoff risks missing very valuable textual information that might be stored in remotely homologous genes. However, using too liberal of a sequence similarity cutoff may result in a flood of references.

One solution is to transfer references only from a fixed number of the most similar well-studied sequences; these are the genes with the most reliable and extensive documentation. In addition we would likely only want to transfer specific references; those references that refer only to the well-studied gene and not to too many other genes are ideal. Simple screens that remove genes with less than a preset number of references and that remove references that refer to more than a preset number of genes can achieve these goals in a straightforward way.

In Chapter 7 we will actually utilize this approach to facilitate gene expression analysis. One practical implementation of this strategy is available online (Tu, Tang et al. 2004).

4.3 Assigning keywords to summarize sequence hits

Given a query sequence of unknown function, one of the primary goals of database sequence searches is to assign an appropriate biological function based on the most similar sequences (Andrade and Valencia 1998; Shatkay, Edwards et al. 2000; Masys, Welsh et al. 2001). This is an area that applies not just to sequence similarity searches, but to any functional genomics query where the result is a group of genes or sequence that may have shared function.

Of course, there are many approaches to this same problem. The general goal is to look for functional words that are used more frequently with the sequences in the resulting query compared to other sequences or the corpus of biology in general.

The simplest strategy is to use weighted word vectors to represent the documents as described in Chapter 3. For all of the genes we can construct a weighted word vector that is created by averaging the weighted word vectors for each of the referring documents. We saw in Chapter 3 that the words with greatest weight in this textual profile were good keywords for an individual gene. We demonstrated this with the fly genes *heartless* and *breathless*. We can extend this strategy by now averaging together the weighted word vectors for all of the sequences obtained from a search.

As an example we look at the BLAST search that we conducted in Chapter 3. We have depicted the results for that search in Plate 4.1. The sequences seem to line up with one particular domain ranging about from positions 700 to 1000 on the *breathless* query sequence. As we mentioned in the previous chapter, this query corresponds to 67 unique *drosophila* genes. To ascertain the function of this region, we can look at the 32 averaged textual profiles of the unique genes with five or more references. We average these weighted textual profiles. The top five weighted words are listed in Table 4.1. These words suggest that this domain may correspond to the tyrosine kinase domain of the protein. A more comprehensive survey would include sequences from many organisms, and not just *drosophila*. If a sequence has groups of hits to different areas, the hit sequences can be grouped according to the corresponding area that they have similarity to in the original sequence; then each group can be used to find keywords that describe each domain specifically.

Table 4.1 *Keywords to describe sequence similarity hits for breathless.*

Word	tf/idf weight
kinase	3.69
tyrosine	2.91
signalling	2.10
kinases	1.78
cell	1.62

This method's effectiveness is contingent on the weighting strategy employed. Term frequency/inverse document frequency weighting schemes emphasize words that are rare in the entire biological corpus. We showed in Chapter 3 that rare words are the most functionally relevant, and likely have the potential to be good keywords. If these words appear relatively more frequently in the set of documents associated with the sequences, then these rare and heavily weighted words become very good keyword candidates.

An alternative strategy is to identify statistical properties of word frequencies, and given a series of sequence hits, assess how the word frequency among those hits compares. In an implementation of this approach, a predefined set of protein families were obtained to characterize the distribution of all vocabulary words (Andrade and Valencia 1998). For each protein sequence, an unweighted word vector is constructed. Then for each word and family of protein sequences a frequency statistic is calculated:

$$F_{i,j} = \frac{W_{i,j}}{S_j}$$

where $W_{i,j}$ is the number of sequences in family j that word i appears in, and S_j is the total number of sequences in that family. Then for each word i, calculate a mean frequency and a frequency standard deviation:

$$\bar{F}_i = \sum_j \frac{F_{i,j}}{N}$$

$$\sigma_i = \sqrt{\frac{\sum_i (F_{i,j} - \bar{F}_i)^2}{(N-1)}}$$

where N is the number of families in the set. Given a new family of protein sequences, the $\bar{F}_i$ and σ_i for each word i are calculated just as above. Given a new family of proteins, the z-score (see Section 2.2) for each word can be calculated:

$$Z_i = \frac{F_{i,j} - \bar{F}_i}{\sigma_i}$$

Words with high z-scores are selected as keywords. The idea is that words that occur with protein families infrequently on average and that have occurred with the observed family with higher than expected frequency are likely high quality keywords. There is no clear consensus at this point about which of these methods or others are optimal.

4.4 Using textual profiles to organize sequence hits

Text can be used in sequence analysis to organize the results of sequence similarity searches (McCallum and Ganesh 2003). Once a sequence similarity search has been completed, the textual records associated with each of the hit sequences can be obtained and converted into document word vectors. As described in Chapter 3, standard statistical procedures such as clustering can be utilized with document vectors and textual profiles. There is a detailed discussion on clustering in Section 2.4. So, these methods can be applied to cluster hit sequences based on their textual profiles as well. The result is that sequence hits are partitioned into groups. Since the textual contents contain functional information, the sequences will be grouped by their functions. McCallum and Ganesh explored hierarchical clustering and k-means clustering of sequence hits based on text. They demonstrated that larger clusters will likely be most relevant to the query sequence. Very small clusters may represent spurious hits. The different clusters may represent different functions of the same query sequence. For example a sequence with multiple functional domains may have similarity to different groups of sequences. Clustering should be able to separate the sequences corresponding to the different functional domains. The keyword strategies described in the previous section could be applied to the clusters separately to identify the relevant functions.

4.5 Using text to help identify remote homology

One goal of sequence similarity searching is to identify remotely homologous genes. When the sequence similarity between two genes is not significant enough to be certain of their evolutionary relationship, investigators must look for other evidence. Since gene function is often preserved through evolution, homologous genes often have similar functions. If the common function between two somewhat similar sequences can be appreciated, one can conclude that an evolutionary relationship is likely.

In practice, biologists will often manually inspect the keywords and references associated with low scoring sequence similarity hits, and assess whether there is possibly a relationship to the query sequence. It may be possible to automate this tedious manual inspection that occurs after a search.

After application of a sequence similarity (i.e. BLAST) algorithm to search for homologous sequences, one strategy is to a filter low sequence similarity hits based on functional similarity using textual profiles. For those sequences calculate the cosine between the weighted word vectors of the query sequence and each of the hits as a textual similarity measure. Sequences with low textual and sequence similarity are considered to not be homologous genes.

One group implemented this approach using PSI-BLAST to obtain sequence hits (MacCallum, Kelley et al. 2000); PSI-BLAST is described in detail in Section 2.3 (Altschul, Madden et al. 1997). Then for sequence documents, authors used text from SWISS-PROT record entries that included keywords, reference titles, and comments. The authors did not extract the abstract text of the references from the PubMed database. The authors then created word vectors for each sequence entry in SWISS-PROT. A weighting scheme, similar to the one we introduced in Chapter 3, was used to increase the weight of rare terms. For a given query sequence, a list of similar sequences and their similarity scores was obtained. In addition, the authors calculated a text-based similarity between the query sequence and the hit sequences. They compared the sequence similarity score versus the text-based similarity score for sequences obtained in sequence queries. In fact, they found that among high scoring sequence hits, text similarity might actually be more indicative of homology than sequence similarity. They propose that a weighted combination of the two scores might be an appropriate way to rank sequences.

4.6 Modifying iterative sequence similarity searches to include text

The methods that we have discussed to this point in the chapter have used scientific text to organize, summarize, or modify the results produced by a sequence similarity search algorithm. Instead of simply summarizing or modifying search results, scientific text can be used as an integral part of sequence similarity searches as well. One strategy would be to implement an iterative algorithm that first seeks similar sequences to the query sequence. Then the algorithm would modify the results based on text similarity to the original sequence. The resulting sequences can be used to create a multiple alignment that can be used to search the sequence database again to update the collection of sequences. This process could be iterated until a stable collection of sequences with both literature and sequence similarity was achieved.

PSI-BLAST is an iterative sequence search algorithm that can be easily modified to include text. Given a query sequence, in the first iteration, a BLAST search obtains significantly similar sequences that are used to create a probabilistic sequence profile. In subsequent iterations that profile is used to search the database and to update the significant sequences (see Figure 4.2). By including more diverse sequences into the query, sensitivity is improved.

However, as PSI-BLAST iterates, it includes a more diverse array of sequences, and the possibility of including a spurious sequence that is not a homolog of the original query sequence increases. Thus, any errors introduced into the profile can be magnified, eventually diluting the signal from the original sequence. This situation has been called "profile drift". In these situations the algorithm fails to converge or converges to an imperfect solution. PSI-BLAST considers only sequence similarity. As we saw in the previous section the scientific literature associated with the sequences can also provide valuable information. For example, suppose a query sequence

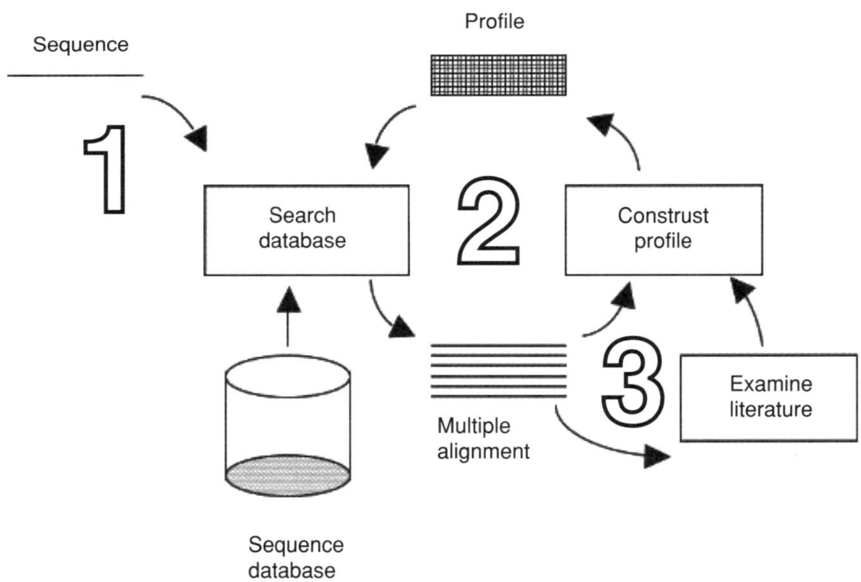

Figure 4.2 *An illustration of PSI-BLAST to include textual information.* A sequence is used in the initial query to BLAST search the database for similar sequences (1), a multiple alignment is then used to construct a profile to search the database again (2). The modification (3) involves screening the sequences that constitute the multiple alignment for literature similarity; the sequences for which the associated literature is least concordant with that of the original sequence used in (1) are eliminated from the construction of the profile.

is similar to many kinase proteins. A reasonable refinement may be to consider only those proteins that are also kinases. During each iteration of the sequence similarity search, sequences can also be examined for textual similarity, and inclusion of this additional information may result in a search that is relatively resistant to contamination. One group demonstrated modest performance improvement by utilizing text in PSI-BLAST (Chang, Raychaudhuri et al. 2001). Their adaptation of PSI-BLAST removes sequences that lack sufficient literature similarity in each iteration.

The method requires the creation of textual profiles for each sequence. For each sequence, the authors utilized the description, comments, and keywords from the SWISS-PROT record. In addition, they utilized the record's references from PubMed; they used the MeSH headings, subheadings, and text from the abstracts records. The authors construct unweighted word vectors for each sequence.

At each iteration, this PSI-BLAST implementation eliminates sequences that have poor literature similarity to the query sequence. After each sequence search iteration, the cosine between the word vector of each significant hit and the query sequence is calculated. The scores are ranked according to their text similarity scores; the sequences with the lowest scoring text similarity are discarded, thereby excluding them from the profile (Figure 4.2). The authors experimented with removing different percentages of sequences.

A limitation of any natural language processing approach to biological problems is that areas for which the appropriate quantity of text is unavailable may be difficult to study. In the context of this work, for example, annotation of newly discovered sequences is challenging since sufficient descriptive text is lacking.

4.7 Evaluating PSI-BLAST modified to include text

Validation of sequence similarity algorithms requires a high quality gold standard. In an ideal case, the gold standard should contain families of homologous proteins. If a query with one of the proteins returns all of the other members of the family that it belongs to, then we feel confident that the method works well. Homologous families should contain sequences that are related by evolution, rather than just by sequence similarity. Since this is difficult to define, one option is to use a definition based on the Structural Classification of Proteins Database (SCOP) (Murzin, Brenner et al. 1995). SCOP is a manually constructed hierarchical categorization of proteins based on structure and function. Since biological relatedness is implied at the superfamily level, one appropriate definition of a homology family is the set of SWISS-PROT sequences that reference structures in the same SCOP superfamily (Lindahl and Elofsson 2000). All SWISS-PROT sequences that map into a single SCOP superfamily can be used as a gold standard for a family.

To validate the modified PSI-BLAST approach described in the previous section, the authors used a set of query sequences that were as divergent as possible from the families they belonged to (Chang, Raychaudhuri et al. 2001). They compared the performance of their modification to the standard PSI-BLAST algorithm.

Figure 4.3 plots their results; it shows a comparison of the performance of PSI-BLAST to the modified PSI-BLAST approaches. "Recall" is the number of homologous sequences surpassing a fixed e-value cutoff divided by the total number of homologous sequences. At a fixed recall, "precision" is the number of homologous sequences detected divided by the total number of sequences detected. The ideal algorithm would maintain 100% precision for all values of recall; that is, it would be a straight line across the

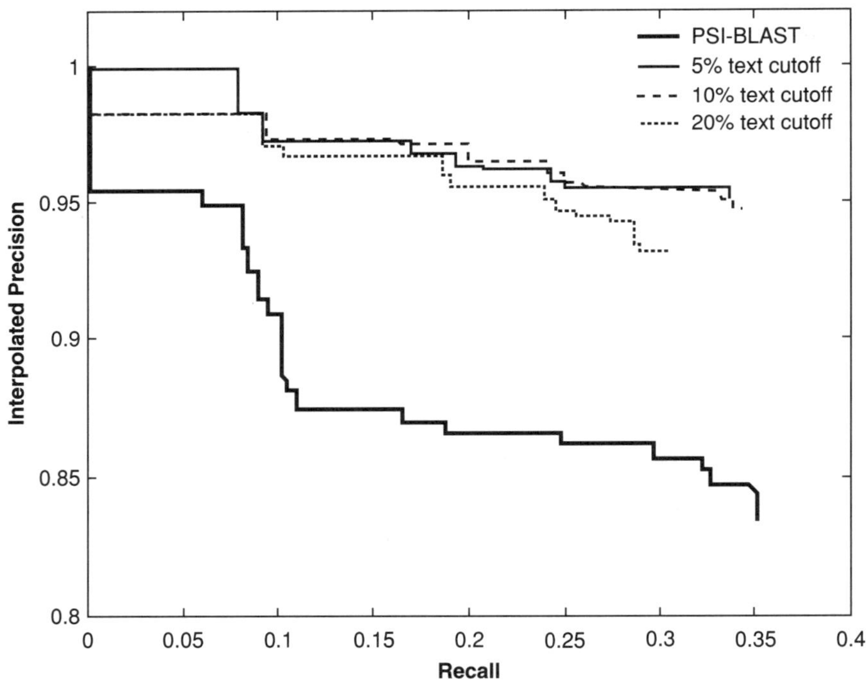

Figure 4.3. *Using text comparison improves homology search results.* Results of homology search for 54 training sequences from different families. Precision was interpolated to insure that the curves were monotonic. The solid bold line represents the unmodified PSI-BLAST algorithm; other lines represent the modified PSI-BLAST algorithm that drops the sequences with the lowest 5%, 10%, and 20% of literature similarity.

top. The modified PSI-BLAST was more precise than the original at any recall. In addition, the precision did not decay as rapidly as recall was increased.

For 46 of the 54 SCOP families that were tested, the outcome was identical for the modified and the unmodified PSI-BLAST. Out of the eight queries remaining, five differed in convergence, while three differed in performance. It is these eight families that account for the differences in the performance plotted in Figure 4.3.

These eight families fall into three categories. The first two families in Table 4.2 converged to poor solutions with standard PSI-BLAST and failed to converge for the modified PSI-BLAST. The next three failed to converge for PSI-BLAST, but converged to reasonably good solutions for the modified PSI-BLAST. The final three converged for both modified and standard PSI-BLAST; the solutions are slightly better for the standard one. In these three cases performance differences can be attributed to a single missed sequence in each family.

Table 4.2 *Comparing PSI-BLAST and a modified version of PSI-BLAST that includes text.* Most of the 54 families have identical performance for both algorithms and are not shown when 10% of the sequences are dropped at each PSI-BLAST iteration based on textual dissimilarity. The other families where there is a performance difference are shown below. "Superfamily" is a SCOP homology family and "Query sequence" is its representative. "Words" is the number of document words associated with the query sequence. "# Seqs" is the number of sequences in the family. The final six columns describe the results of a search with the query sequence. Here, precision and recall were calculated for each individual family using all the results from the homology search.

				Convergence		Precision		Recall	
Superfamily	Query sequence	Words	# Seqs	PSI-BLAST	Text 10%	PSI-BLAST	Text 10%	PSI-BLAST	Text 10%
EGF/Laminin	C1R_HUMAN	1661	5	yes	no	0.11	N/A	0.8	N/A
Acid proteases	POL_HV2RO	1271	22	yes	no	0.6	N/A	0.27	N/A
PLP-dependent transferases	GLYC_RABIT	1052	21	no	yes	N/A	1	N/A	0.1
Thioredoxin-like	CAQS_RABIT	1516	13	no	yes	N/A	1	N/A	0.38
Glucocorticoid receptor-like (DNA-binding domain)	CYSR_CHICK	1738	10	no	yes	N/A	0.8	N/A	0.4
EF-hand	SCP_NERDI	963	31	yes	yes	0.92	0.92	0.74	0.71
Glycosyl-transferases	CHLY_HEVBR	1007	20	yes	yes	1	1	0.2	0.15
Snake toxin-like	CD59_HUMAN	2435	23	yes	yes	1	1	0.13	0.09

For the "EGF/Laminin" and "Acid proteases" families the standard PSI-BLAST converged upon incorrect answers, indicating that drift occurred. Modifying PSI-BLAST to include text slowed the drift and prevented convergence. These families were challenging because non-homologous sequences had high similarity to family sequences. Literature similarity checking added an additional constraint against including erroneous sequences.

For the protein family "Thioredoxin-like", the PSI-BLAST homology search with the "CAQS-RABIT" test sequence failed to converge. The modified PSI-BLAST that utilized literature similarity did converge on a precise solution; it correctly detected five sequences. In this case, removing sequences with low literature similarity prevented profile drift and allowed the search to converge on a correct solution.

The literature similarity constraint made little difference in the performance of PSI-BLAST in the majority of the families. So the approach of including scientific text presented here certainly does not hurt performance, and in special cases, significantly improves performance.

4.8 Combining sequence and text together

One promising area is the utilization of information contained in text with sequence information to determine the function of a protein. Classification algorithms can be trained with proteins of known function on features from both text and sequence. These classifiers can then be used to make functional predictions on uncharacterized proteins. This strategy has been applied with success to predicting the subcellular location of proteins.

It is well accepted that protein sequences provide biological information that can be used to make reasonably accurate subcellular location predictions (Feng 2002). Addition of textual information can further improve predictions by including documented known aspects about the protein. Several groups have experimented with machine learning approaches that combine information about sequences as well as textual information to make predictions about subcellular localization of the cell (Stapley, Kelley et al. 2002; Eskin and Agichtein 2004; Lu, Szafron et al. 2004).

In one of the approaches the authors took unknown sequences and conducted BLAST sequence similarity searches to find the three most similar SWISS-PROT sequences for each one (Lu, Szafron et al. 2004). These sequences are presumed to be homologous sequences. The text of the keyword field, subcellular localization subfield of the comment field, and InterPro family number from the database source field were extracted from the three protein records. These textual phrases were then used as features in a classifier that could predict subcellular location of the protein. The investigators experimented with neural networks, naïve Bayes, nearest neighbor, and support vector machines. They found that all of the classification methods achieve reasonably high accuracy in predicting cellular location of the protein. This approach is similar to summarizing the results of a protein sequence similarity search as described above.

In a second study investigators used text in their analysis to two separate ends (Eskin and Agichtein 2004). First, the authors noted that there was a lack of properly annotated sequences where the subcellular localizations are available to them. They used classification approaches on text in SWISS-PROT records alone to assign subcellular locations to as many proteins as possible with accuracy; this augmented their set of sequences with known subcellular localization. This is akin to assigning a function to a gene using scientific text alone (see Chapter 8 for a full discussion on functional assignment).

The purpose of augmenting this set was to have more training examples for classifiers, and ensure more accurate predictions than would be possible with a small number of training examples. After augmenting their training set, the authors went on to define for each sequence two feature vectors. One contained counts of words associated with the sequence's SWISS-PROT record. The second vector contained sequence information about the protein. They then used a joint classifier that classified proteins on the basis of both features of their sequence as well as text in the SWISS-PROT records. They found that the combined approach achieved better precision and recall rather than just classifying on sequence or text alone.

The approach of classifying sequences based on both their sequence features and their textual features is a very new and exciting area. We expect in the coming years there will be more examples like these to determine protein function automatically.

References

Altschul, S. F., T. L. Madden, et al. (1997). "Gapped BLAST and PSI-BLAST: a new generation of protein database search programs." *Nucleic Acids Res.* **25**(17): 3389–402.

Andrade, M. A. and A. Valencia (1998). "Automatic extraction of keywords from scientific text: application to the knowledge domain of protein families." *Bioinformatics* **14**(7): 600–7.

Boeckmann, B., A. Bairoch, et al. (2003). "The SWISS-PROT protein knowledgebase and its supplement TrEMBL in 2003." *Nucleic Acids Res.* **31**(1): 365–70.

Chang, J. T., S. Raychaudhuri, et al. (2001). "Including biological literature improves homology search." *Pac. Symp. Biocomput*: 374–83.

Eskin, E. and E. Agichtein (2004). "Combining text mining and sequence analysis to discover protein functional regions." *Pac. Symp. Biocomput*: 288–99.

Feng, Z. P. (2002). "An overview on predicting the subcellular location of a protein." *In Silico Biol.* **2**(3): 291–303.

Lindahl, E. and A. Elofsson (2000). "Identification of related proteins on family, superfamily and fold level." *J. Mol. Biol.* **295**(3): 613–25.

Lu, Z., D. Szafron, et al. (2004). "Predicting subcellular localization of proteins using machine-learned classifiers." *Bioinformatics.* **20**(4): 547–56.

MacCallum, R. M., L. A. Kelley, et al. (2000). "SAWTED: structure assignment with text description–enhanced detection of remote homologues with automated SWISS-PROT annotation comparisons." *Bioinformatics* **16**(2): 125–9.

Masys, D. R., J. B. Welsh, et al. (2001). "Use of keyword hierarchies to interpret gene expression patterns." *Bioinformatics* **17**(4): 319–26.

McCallum, J. and S. Ganesh (2003). "Text mining of DNA sequence homology searches." *Appl. Bioinformatics* **2**(3 Suppl): S59–63.

Murzin, A. G., S. E. Brenner, et al. (1995). "SCOP: a structural classification of proteins database for the investigation of sequences and structures." *J. Mol. Biol.* **247**(4): 536–40.

Shatkay, H., S. Edwards, et al. (2000). "Genes, themes and microarrays: using information retrieval for large-scale gene analysis." *Proc. Int. Conf. Intell. Syst. Mol. Biol.* **8**: 317–28.

Stapley, B. J., L. A. Kelley, et al. (2002). "Predicting the sub-cellular location of proteins from text using support vector machines." *Pac. Symp. Biocomput*: 374–85.

Tu, Q., H. Tang, et al. (2004). "MedBlast: searching articles related to a biological sequence." *Bioinformatics* **20**(1): 75–7.

5 Text-based analysis of a single series of gene expression measurements

In this chapter we begin to address the issue of the analysis of gene expression data with the scientific literature. Here we describe methods for the analysis of a single experiment—one where a single expression measurement has been made for many genes within the same organism. In Chapter 7 we will address the analysis of larger data sets with multiple expression measurements for each of the genes; the questions that occur in that setting are often more complex and utilization of scientific text in that setting can be more useful. But focusing on a single series of expression measurements is an effective starting point in understanding the scientific literature and how it can be used with experimental data. The lessons here can be applied to a wide array of genomic assays besides gene arrays. These methods can be applied to any assay that assigns a single value to each gene In addition, many investigators generate single-condition expression data sets, and these methods are widely applicable.

One of the great difficulties in analyzing a single expression series is that context is lacking. That is, we have a large set of isolated measurements. Each measurement corresponds to the log of the relative ratio of a single gene's expression in an experimental condition compared to its expression in a control condition. These measurements represent a single snapshot of a cell's physiologic status. One of the great challenges is sorting out the physiologically important expression changes compared to random experimental and physiologic aberrations and fluctuations. Gene expression measurements are subject to a great amount of noise and distinguishing true positives from genes that are not truly induced or repressed is a great challenge. Typically, investigators use their knowledge of biology to prioritize likely positives. In this chapter we argue that text-mining approaches can be used to help prioritize these genes instead.

Another equally important challenge is to discern broadly what biological functions are active in a given experiment. A quick list of keywords culled from the literature could offer a rapid view of the broad processes that are induced or repressed in a given experiment.

The basic concepts of this chapter are highlighted in the frame box. For this chapter we introduce a phosphate metabolism data set; we use this data set to demonstrate these key concepts. We begin with a discussion of the confounding factors in gene expression measurements and how they result in many false positives and negatives. We then talk about the basic statistical properties of a gene expression data set. We motivate and introduce neighbor expression information (NEI) scoring, a text-based information theoretic score. It has applications at recognizing true and false positive genes in an expression series. It can also be used to roughly gauge the quality of an experiment. Finally we introduce methods to assign keywords to an expression series from text about genes.

5.1 Pitfalls of gene expression analysis: noise

In looking at a single gene expression experiment the goal is to distinguish genes that are aberrantly expressed under some physiological stress or environment change. The difficulty is that in the assessment of some several thousand genes the likelihood of a certain number of false positives, when the true positives may only be a handful of genes, is enormous. For example, consider a gene expression assay of 1000 genes where 30 are truly responsive to the physiological stimuli being assessed. Assume our expression assay is 90% sensitive and 95% specific. In this case we will discover 27 of the 30 truly responsive genes. But, on the other hand, we will also incorrectly identify 49 genes that are not actually responding to the physiologic stimuli. So 64% of the 76 genes detected by this assay are entirely spurious. In practice distinguishing true positives and false positives requires careful manual assessment of the results, experimental followup, and experimental replication.

The difficulty is the many sources of noise involved in gene expression assays; these have been explored in the literature (Lee, Kuo et al. 2000; Novak, Sladek et al. 2002; Tu, Stolovitzky et al. 2002).

1) Noise sources in gene expression measurement
2) Statistical properties of an expression data series
3) Neighbor expression information method
    a) Separating true and false positives
    b) Assessing experiment quality
4) Finding keywords to describe an experiment

Creation of high quality gene expression data requires exposing cells to both an experimental and control condition; the mRNA from these cells is then harvested, amplified and labeled and hybridized to a gene expression array; see Section 2.4 for a more detailed discussion. The log base 2 of the ratio of the signal between the experimental and control condition is reported as the expression for each of the genes. Sources of variability and noise include technical sources, physiologic sources, sampling sources in the lab, as well as sources in the actual array as well.

Some of physiologic variability includes differences in the way the biological samples may have been prepared. For example, media conditions that cells may have been cultured in, or variations in presumed important lab conditions could account for significant gene expression influences. Sampling sources include expression variability that might occur while harvesting a particular tissue; contaminant tissue or heterogeneous cells might be inadvertently included and may alter the measured expression profile. This is a frequent issue in medical applications using pathlogical samples. Gene expression array analysis is also plagued by a host of laboratory technical issues. These would include the quality of the work used to isolate the mRNA, the quality of the mRNA samples, and the efficiency of reverse transcribing and labeling the mRNA with fluorescently dyed nucleotides.

In addition to these issues, investigators are familiar with many chip or array noise sources. For example, expression measurement variability can be attributed to the quality of the probe on the slide, the ability of the probe to bind labeled DNA well, and even the position of the spot on the slide. The probe on the slide may have cross-reactivity to other genes and bind other labeled nucleotide products than the ones that it is supposed to, and hybridization may be altered. Genes that have low expression in either the control or experimental condition are particularly susceptible as the ratios between the two conditions can be dramatically affected by these small errors. Specific regions of the glass slide may be smeared or damaged, affecting the expression of a large number of genes.

These error sources account for the differences between true physiologic gene expression and the gene expression that is measured by an expression assay. When looking at a single experiment and trying to appreciate the genes with truly altered gene expression we must realize that a large number of the genes that may appear induced or repressed may appear that way only because of noise. The challenge is to separate the genes whose alteration in expression is biologically relevant versus the aberrant measurements.

Of course, the inclusion of external information to bias our analysis helps to distinguish biologically important gene expression changes from aberrant ones. For example, additional gene expression conditions may suggest the true positives; these will have increased gene expression under other similar physiologic conditions. Genome sequence information can be suggestive also. For example, if a set of induced genes shares a common promoter site then that may suggest that those genes correspond to true altered expression. Here we discuss the possibility of using the scientific literature; our hypothesis is that the likelihood that a gene has truly altered gene expression under a physiologic condition is related to whether or not genes that are functionally similar also have altered expression under the same condition. These functionally similar genes can be identified with the scientific literature.

5.2 Phosphate metabolism: an example

In this chapter we will focus on a series of experiments conducted by Saldhana and colleagues that explored the gene expression of yeast under nutritionally deficient circumstances (Saldanha, Brauer et al. 2004). The authors investigated the changes in gene expression that occurred when different metabolites were made available to yeast cultures in limiting quantity. They explored whether and at what point the limited availability activated particular metabolic pathways, or stress responses. The data set included a total of 124 experiments, and measured the expression of some 5104 genes. We will focus on six replicates of what the authors referred to as PU conditions; in this series the organisms were placed in phosphate deprived media in the experimental conditions, whereas in the control conditions uracil deprived media were used instead. So genes with higher expression values have high expression under phosphate deprivation as compared to uracil deprivation. The histogram of gene expression ratios are displayed in Figure 5.1.

We chose this data set for several reasons. The first is that it is a yeast data set and the literature in yeast is relatively well organized, and provides a good starting point to illustrate literature-based approaches. The second is that the data set includes six replicates of the same experiment. Averaging the expression values of the six data sets together mitigates some of the noise. For example, error associated with laboratory protocol may be reduced by averaging replicates. This affords us something closer to a higher quality experiment. For the purposes of demonstration this is ideal. Finally, the experiment is relatively easy to understand. Since these yeast cultures were phosphate starved, it is very reasonable to expect that genes related to phosphate metabolism should be aberrantly expressed. These factors make this data set well suited for demonstrations, and avoid some of the ambiguity that is inherent to most gene expression array analysis. The raw data was obtained from the Stanford Microarray Database (SMD) (Ball, Awad et al. 2005).

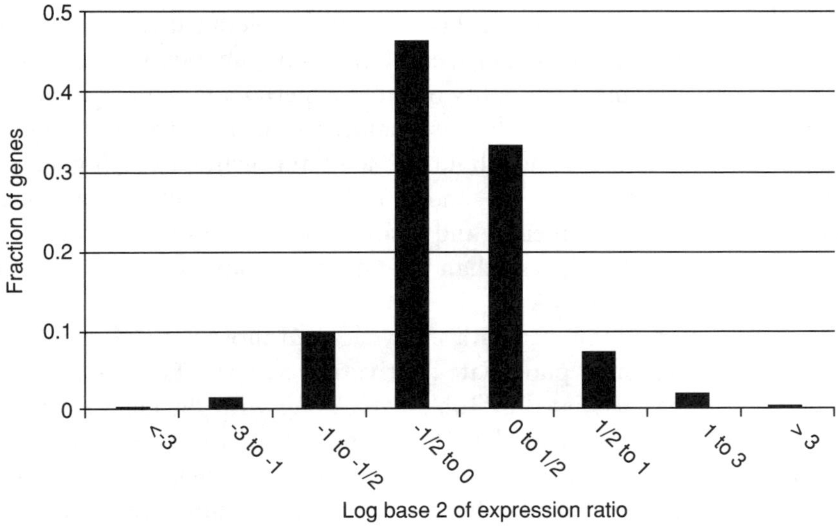

Figure 5.1 *Histogram of phosphate–uracil experiment expression log ratios.* Expression log ratios from all six experiments are averaged together and a histogram is plotted.

Of the 5104 genes analyzed in this data set, we were able to identify 3611 unique genes that had at least one article. Article references were obtained from the Saccharomyces Genome Database; PubMed abstracts were used. Varation in gene nomenclature may have prevented some genes from being associated with references. The mean number of articles per gene was 14.3, and the median number of articles per gene is 5. The highest number of article references per gene was 443. A total of 74 genes had over 100 references. This data set referenced a total of 21,213 PubMed abstracts. Each article referred to a mean of 2.43 genes and a median of three genes. The largest number of genes referred by a single abstract was 132. A total of seven articles referred to more than 50 genes, and 234 referred to over ten genes. The distributions of genes per article and articles per gene are extremely skewed distributions; this issue is addressed in greater detail Chapter 1.

In the remainder of this chapter we demonstrate how literature-based approaches can be used to distinguish true positive genes that are expressed in this data set from the false positives using a literature based scoring system. In addition, we demonstrate one method to quickly assign keywords that broadly describe the experiment.

5.3 The top fifteen genes

The distribution of gene expression after averaging the six similar replicates is depicted in Figure 5.1. We focus on the average of the six experiments rather than an individual one in this section; the averaged experiment is a surrogate for a high quality experiment. Later in this chapter we will look at individual experiments; the quality of those experiments is more typical. It can be seen immediately that the vast majority of genes have very minimal changes in gene expression. About 80% of the genes have log base 2 expression ratios between −0.5 and 0.5. This corresponds to expression ratios between the experimental and control condition ranging from 0.7 to 1.4. On the other hand, less than 1% of genes have expression ratios exceeding 8 or less than 1/8.

To begin exploring this rich data set, we look at those genes that have the greatest induction under phosphate deprivation compared to uracil deprivation. These genes are listed in Table 5.1. Along with the genes, we have listed a short description of their functions. All of these genes are very significantly expressed in this data. Do they all make sense in the context of the experiment? Certainly the genes that are intimately involved in phosphate metabolism are consistent with our understanding of the data. So genes in rows 1–3 and 5 are almost certainly true positives. A number of the genes are involved in vacuole fusion; as we can see with gene VTC3, there seems to be a connection between vacuolar transport and phosphate

Table 5.1 *The highest expressed genes in the PU experiments.* We have listed the top 15 expressed genes in order. Also listed is the systematic name, the gene function, the average expression log ratio, and the corresponding ratio. Many genes are related to phosphate metabolism.

Gene	Systematic name	Function	Expression (log ratio)	Expression (ratio)
1 PHO11	YAR071W	Phosphate metabolism	6.3	78.8
2 PHO12	YHR215W	Acid phosphatase	5.6	48.5
3 PHO5	YBR093C	Phosphate metabolism	5.1	34.3
4 un-named	YOL155C	Cell wall Organization	4.98	31.6
5 PHO3	YBR092C	Acid phosphatase involved in thiamine transport	4.16	17.9
6 VTC3	YPL019C	Phosphate metabolism Vacuole fusion (non-autophagic)	3.72	13.2
7 VTC1	YER072W	Vacuole fusion (non-autophagic)	3.49	11.2
8 BAP3	YDR046C	Amino acid transport	3.43	10.8
9 un-named	YAR068W		3.39	10.5
10 ARO9	YHR137W	Aromatic amino acid family metabolism	2.92	7.6
11 SSU1	YPL092W	Sulfite transport	2.65	6.3
12 SUL1	YBR294W	Sulfate transport	2.52	5.7
13 HXT2	YMR011W	Hexose transport	2.52	5.7
14 un-named	YIL169C		2.41	5.3
15 VTC4	YJL012C	Vacuole fusion (non-autophagic)	2.33	5.0

metabolism. So these genes might be reasonable candidates. The remaining genes are involved in diverse processes or unknown processes that do not have any obvious link to phosphate metabolism.

5.4 Distinguishing true positives from false positives with a literature-based approach

Our goal is to use the literature to help distinguish the true positive genes from the false positive genes. The degree of noise in gene expression is great enough that it is a challenge to set a concrete expression threshold and identify induced genes with a sufficiently high degree of sensitivity and specificity. The solution is to look at the genes themselves and to understand if it is reasonable to expect those genes to be induced or not. In a best-case scenario, such as the one we are presenting with our example data, we have an excellent idea what the implications of the physiologic stress or condition is and what types of genes we would expect to be induced.

An alternative approach is to assume that genes that are truly induced in a physiologic condition are not likely induced in isolation. A physiologic stress that induces the expression of a gene likely affects other related genes involved in similar biological processes.

As we have suggested in Chapter 3, the literature can be used to identify functionally similar genes. The approach we take here is that for each gene that might be significantly expressed, we identify likely functionally similar genes and examine what their expression is. If their expression is also significantly affected, then we are likely looking at a gene that is truly induced in this physiologic condition.

For each gene we create a normalized and weighted word vector using the references to that gene as detailed in Chapter 3. Then we calculate distances between that gene and all of the others in the literature space using the cosine distance metric. We assume that many of the most similar genes share some function.

As an example consider the most expressed gene in this data set, PHO11. In Table 5.2 we have listed the 20 most similar genes to PHO11 based on word vector similarity. In addition we list the log expression ratios for those similar genes. Many of these genes are either very highly induced or repressed, that is, they either have a very positive or negative log expression ratio. Only nine out of the 20 genes have an expression ratio between −0.5 and 0.5. If this were a random distribution of genes we would expect about 80% or 16 genes to have log ratios between −0.5 and 0.5. We see imme-

Table 5.2 *Neighbors of PHO11.* Using the cosine vector metric we calculated the most similar genes to PHO11. We also list their gene expression values.

Neighbors of PHO11	Text-based cosine similarity	Gene expression
PHO5	0.54	5.12
PHO3	0.51	4.16
PHO12	0.46	5.65
PHO4	0.44	−0.23
HIS3	0.43	0.63
PHO81	0.43	0.82
PHO80	0.43	0.53
URA3	0.43	−1.35
LEU2	0.42	−0.69
PHO2	0.42	−0.85
TRP1	0.41	0.05
HIS4	0.39	0.46
LYS2	0.39	0.36
GCN4	0.36	0.14
ABF1	0.36	0.08
LEU1	0.36	−0.4
FLO8	0.35	0.08
ADH2	0.35	−0.12
URA4	0.35	−1.48
URA1	0.35	−3.7

diately that many of the functional neighbors selected using the scientific literature have aberrant gene expression. We might then assume that PHO11 is significantly affected in this condition because it has increased expression on its own, and in addition many genes that have similar function to it also seem to be affected in a non-random way by the condition as well. The next step will be to quantify the non-randomness of the expression of neighbor genes.

5.5 Neighbor expression information

In this section we will present the neighbor expression information (NEI) scoring system; it is a mathematical method that is effective in suggesting whether genes are truly affected by the experimental condition.

For each gene we identify n functional neighbors using similarity in the scientific literature. We average weighted word vectors for all the articles for each gene to define a word vector for each gene. Then we calculate distances between articles using the cosine metric:

$$\frac{x \times y^{\mathrm{T}}}{\sqrt{\| x \| \| y \|}}$$

where x and y are the *tf/idf* averaged weighted word vectors for the two genes. Then, for each gene, we choose the n genes as functional neighbors that have the highest cosine values. These genes have articles that use similar words as each other, and are likely similar in function.

We recognize that while many of these functional neighbor genes may have true functional relationships with the gene, some may not. If the gene is significantly expressed, other neighbor genes with the same function should be perturbed by the same experimental condition; some neighbor genes with dissimilar function may not be affected by the condition at all. In any case, the expression ratio distribution of the functional neighbor genes should be dissimilar to the background distribution of gene expression. We have plotted the expression ratio distribution of the 20 PHO11 neighbor genes alongside the distribution of expression for all genes in this experiment for the sake of comparison in Figure 5.2.

To compare the distribution of expression ratios for functional neighbors to the background we use KL- diveregence, discussed in greater detail in Section 2.2. The KL- divergence is a measure of how inappropriate a background distribution p is at explaining an observed distribution q. In this case, p is the background expression distribution of expression ratios

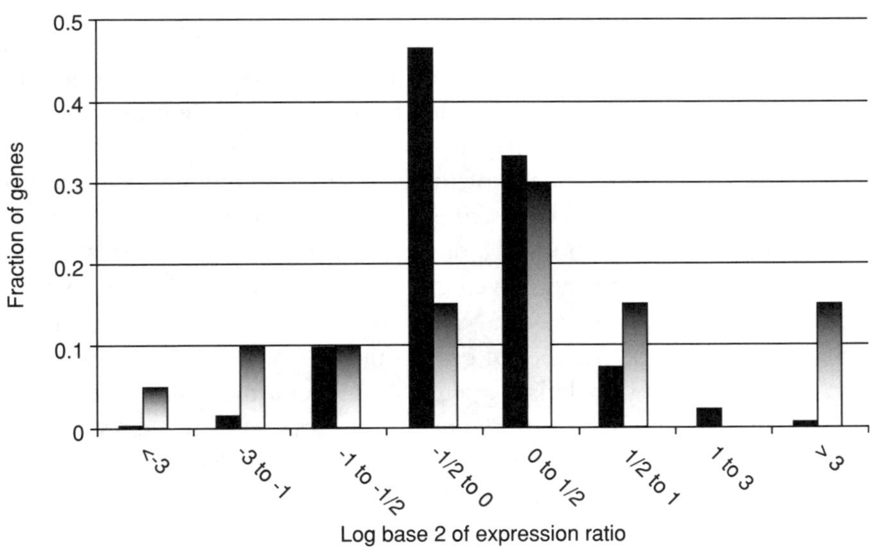

Figure 5.2 *Histogram of PHO11 neighbor expression.* Here we have plotted a histogram of the expression log ratios of the neighbors of PHO11 listed in Table 5.2 with light grey bars. For comparison we have included the histogram of gene expression of all genes for comparison in black; this is identical to the plot in Figure 5.1.

depicted in Figure 5.1. On the other hand, q is the distribution of expression ratios for the genes that are functional neighbors. We calculate:

$$D(q \parallel p) = \sum_i q_i \log_2 \left(\frac{q_i}{p_i} \right)$$

If the expression of the functional neighbor genes is random, then the distribution q should look similar to the distribution p. If the gene is part of a process not affected by the physiological condition, the expression of its functional neighbor genes will likely be distributed randomly, and the divergence will be close to zero. On the other hand, if the gene has functional neighbors that are affected by the process, the distribution q will be enriched in high and low expression genes, and the divergence between the two distributions will be large. So we assume, if we find the divergence to be large, that the gene is likely to be involved in the response to the condition.

So for each gene we calculate the KL divergence of the expression ratios of its neighbors to quantify how likely it is that it is directly involved in the process. We will refer to this number as the neighbor expression information (NEI) score of a gene.

5.6 Application to phosphate metabolism data set

To assess the effectiveness of this measure we apply it to the data set of the six averaged phosphate deficient conditions. We looked at $n = 5$ functional neighbors for each gene. The functional neighbors were identified using the cosine vector distances between gene word vectors as described and the NEI scores were calculated for each of the genes by looking at the expression values of those five genes. The distribution of NEI scores that are obtained is depicted in Figure 5.3. About 5% of genes have an NEI score greater than 1.8, and about 10% of genes have NEI scores greater than 1.4. So focusing our attention on only genes with NEI scores greater than NEI scores of 1.8 or greater reduces the number of genes that we are seriously investigating from 3611 to about 181. This is a much more palatable number of genes to go through. The genes that are the most interesting are the genes with the greatest changes in expression that also have high NEI scores. The NEI scores offer an independent means of evaluating genes with large induction or repression in expression.

Our prediction is that these are the genes that are most relevant to the experiment, and least likely to be false positives. The NEI score is calculated on the basis of literature-based neighbors and their gene expression; it does not consider the expression of the gene itself.

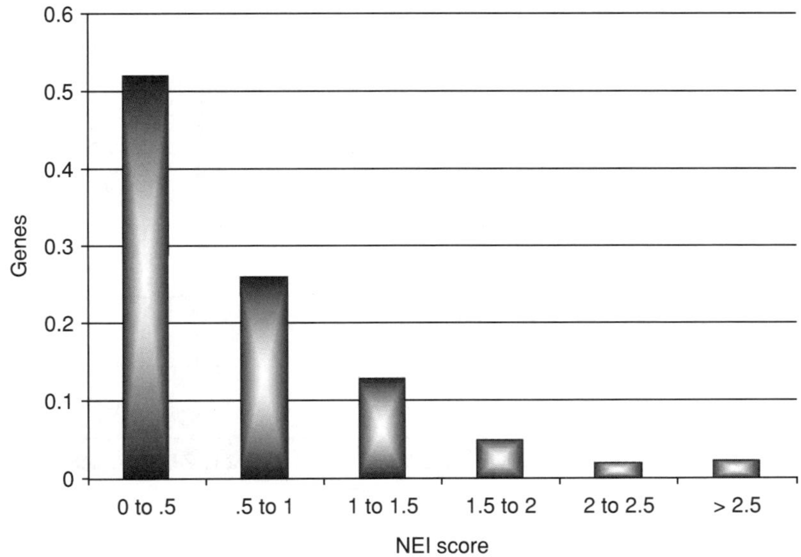

Figure 5.3 *Histogram of NEI scores.*

To validate the NEI scores we see how well it correlates with the expression in this data set. The expression data set being used here is a high quality average of six replicated experiments. Compared to any individual experiments, there should be minimal noise. So in this case, we expect many of the induced and repressed genes to represent true physiological responses. If the high NEI genes are in fact the most relevant to the experiment, it would make sense that genes with higher NEI scores should have, on average, the greatest changes in gene expression. Most of the true positives should have large changes in gene expression and high NEI scores. On the other hand, we would expect few of the genes with minimal changes in gene expression to be involved in the physiologic process and therefore we expect most to have low NEI scores.

In Figure 5.4 we have plotted the absolute log expression ratio as a function of NEI scores. On average, it is apparent the genes with lower NEI scores have minimal changes in gene expression, while those with higher NEI scores have greater changes in expression. Similarly, since the genes that have extremely high or extremely low gene expression values are more likely to be genes that are truly involved in the physiologic process, these genes should have higher NEI scores. Genes that have relatively unaffected gene expression values should likely be uninvolved in the process; these should have relatively low scores on average. In Figure 5.5 we have plotted the median and mean NEI scores for different levels of gene expression. It is immediately apparent that the extremes of gene expression have the mean and median highest NEI scores. In Figure 5.6 we show the

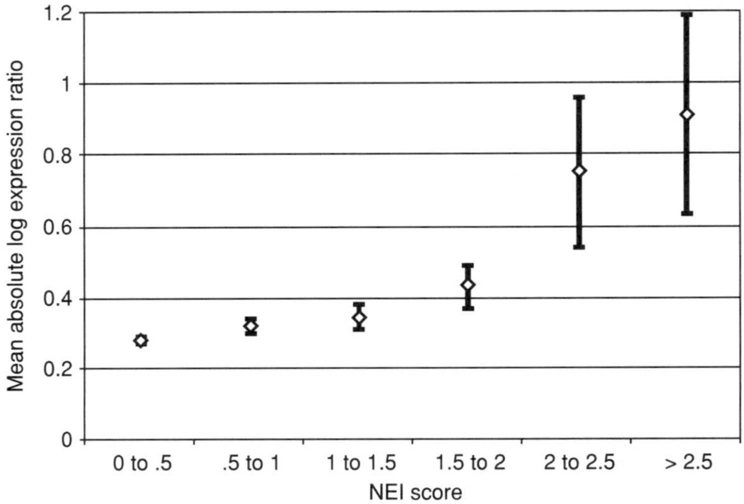

Figure 5.4 *Plot of gene expression versus NEI scores.* Genes are categorized by their NEI scores. The absolute value of the log expression ratio of genes with NEI scores in each range are averaged and plotted. Included are also 95% confidence intervals. Error bars are larger for genes with high NEI scores since there are fewer genes. Genes with high NEI scores have significantly higher gene expression scores than genes with low NEI scores.

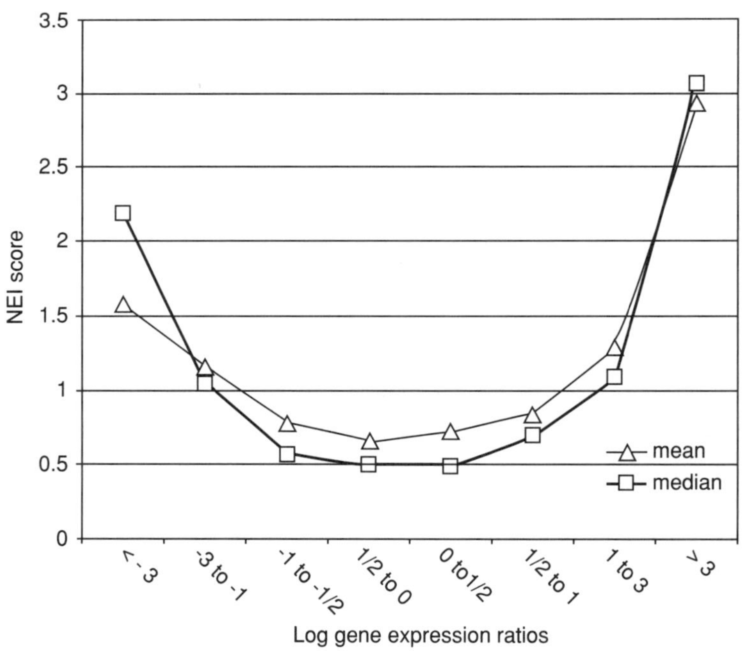

Figure 5.5 *NEI score as a function of log gene expression ratios.* Here all genes are grouped by their gene expression. We plot the mean and median NEI score in each category.

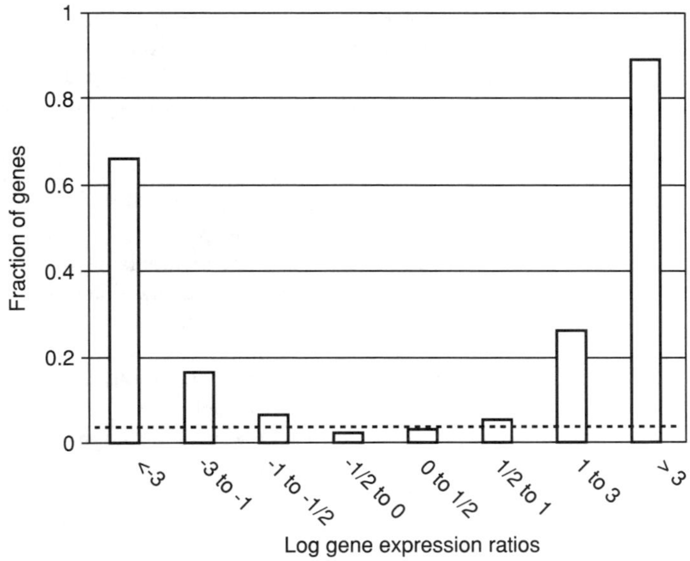

Figure 5.6 *Fraction of genes with high NEI scores as a function of expression ratio.* For each range of gene expression we plot the fraction of genes with NEI scores greater than 2. A horizontal dotted line drawn at 4% indicates the total fraction of genes in the data set with scores greater than 2.

percentage of genes that have NEI scores greater than 2 for different levels of expression. Recall that altogether about 4% of genes have NEI scores greater than 2 in this data set.

Now let us turn our attention to the 15 highest scoring genes in the data set. These were the same genes listed in Table 5.1. We re-list these genes along with their NEI score in Table 5.3. The genes involved in vacuole fusion and phosphate metabolism all have high NEI scores, except for PHO5. These are genes that are almost certainly involved in the condition being tested, phosphate deprivation. On the other hand, the genes ARO9, SSU, and SUL1 are almost certainly false positives, and they have NEI scores that are less than one. The three unnamed and uncharacterized genes are difficult to appreciate; they have very few articles written about them but seem to receive high NEI scores, suggesting they might be true positives. Whether this is in fact the case might be difficult to assess at the present time in an objective sense as there is limited knowledge about these genes.

The NEI scores offer a second way to corroborate the involvement of a gene outside of the expression data itself, and the possibility to distinguish the true positives from the false positives. The NEI scores offer the possibility to go down the list to other induced genes, with lower expression ratios and distinguish which of those are true positives and false positives.

Table 5.3 *NEI scores for the top 15 expressed genes.* In this table we list the same 15 genes as in Table 5.1. We also list the NEI scores with respect to the same gene expression data set, and the number of articles available for that gene. In addition the gene function is listed.

Gene	NEI score	#Articles	Function
PHO11	4.8	14	Phosphate metabolism
PHO12	2.88	3	Acid phosphatase
PHO5	0.81	168	Phophate metabolism
unnamed	3.05	2	Cell wall organization
PHO3	3.05	21	Acid phosphatase involved in thiamine transport
VTC3	3.08	3	Phosphate metabolism, vacuole fusion (non-autophagic)
VTC1	3.08	7	Vacuole fusion (non-autophagic)
BAP3	3.53	13	Amino acid transport
unnamed	2.14	1	
ARO9	0.68	4	Aromatic amino acid family metabolism
SSU1	0.57	8	Sulfite transport
SUL1	0.27	7	Sulfate transport
HXT2	2.02	36	Hexose transport
unnamed	1.78	2	
VTC4	3.12	4	Vacuole fusion (non-autophagic)

5.7 Recognizing high induction false positives with literature-based scores

The example that we have been focusing on so far is a high quality experiment. It is a fabricated experiment that is the average of six different experiments. In this section we will look at poorer quality individual experiments. In these cases, the NEI scores become very valuable, as there are many more false positives.

Here we consider arbitrarily the first of the six phosphate deprivation experiments. The NEI scores of highly induced and repressed genes are in general lower in this data set. The median NEI score for the 15 most induced and repressed genes in this first experiment is 0.99 and 0.72; this compares to 2.9 and 1.5 in the averaged data set (see Table 5.4). In this data set 5% of the

Table 5.4 *NEI scores for each of the individual experiments.* Median NEI scores for the top 15 and bottom 15 genes in each individual experiment are listed in this table. Also the NEI scores for the extreme genes are listed for the averaged experiment and for a randomized experiment.

	exp 1	exp 2	Exp 3	exp 4	exp 5	exp 6	avg	rand
15 most induced genes	0.99	1.56	1.72	2.24	2.5	2.6	2.88	0.49
15 most repressed genes	0.72	0.61	0.83	0.9	1.3	1.44	1.5	0.43
all genes	0.76	0.75	0.62	0.71	0.56	0.65	0.49	0.49

genes have NEI scores above 1.6 and the median NEI score is 0.76. In general NEI scores are lower for the extreme genes. Since the data are much noisier, many more inappropriate genes are highly expressed or induced while many of the genes that respond to the condition have their responses masked in the noise. The inappropriate genes are the false positives, most of which have low NEI scores. In addition, the NEI scores for the true positive genes can be somewhat lower as well, since the expression change of their literature-based functional neighbor genes may be inappropriately lower due to noise. In this example we focus on the most induced genes, and examine how well NEI distinguishes likely false positives from true positives.

In Table 5.5 we have listed the top 15 induced genes in that experiment and their NEI scores, and a short description of their biological functions. Many of the genes have larger expression ratios than in the averaged experiment, but larger expression ratios do not necessarily imply greater biological significance. Only five of these genes are the same as the top induced genes in the averaged data set. Most of these genes have relatively high NEI scores; all five have scores greater than 1. Most of the genes other than those five have no obvious connection to phosphate metabolism. In

Table 5.5 *NEI scores for the top 15 expressed genes in the first experiment.* In this table we list the top 15 expressed genes. We list the gene names in the first and second columns, the gene function in the third column, the log expression ratio in the fourth column, and the NEI Score in the fifth column. Asterisks indicate the genes that were among the top 15 induced in the averaged expession series (see table 5.1).

Systematic name	Name	Function	Log exp ratio	NEI score
YOL155C	*un-named	Cell wall organization	6.7	1.02
YAR071W	*PHO11	Phosphate metabolism	6.3	2.78
YHR215W	*PHO12	Acid phosphatase	5.8	1.89
YLR142W	PUT1	Glutamate biosynthesis	3.7	0.6
YBR150C	TBS1		3.5	0.53
YDR080W	VPS41	Protein transport	3.3	1.24
YLR089C	ALT1	Transaminase	3.2	0.97
YOL038W	PRE6	Ubiquitin dependent protein catabolism	3.2	0.53
YJR148W	BAT2	Branched chain family amino acid biosynthesis	3.1	1.02
YBR092C	*PHO3	Acid phosphatase involved in thiamine transport	3.1	1.73
YAR068W	*un-named		3.1	1.59
YNL333W	SNZ2	Thiamine biosynthesis	3	0.7
YDR017C	KCS1	Response to stress	2.9	0.99
YMR145C	NDE1	Ethanol fermentation	2.7	0.59
YER056C	FCY2	Purine transport	2.7	0.72

addition they have relatively low NEI scores; eight of the ten have NEI scores less than 1. The functions seem rather like a motley collection of unrelated biological functions, and are likely false positives despite their high gene expression. In this context the NEI scores are effective at separating the likely true positives from the false positive. In this case setting an NEI threshold of one would eliminate eight false positives and select all five presumed true positives.

5.8 Recognizing low induction false positives

Another way to demonstrate the value of the NEI scoring system is to focus on genes with even minimal changes in expression values rather than extreme changes. For these genes, it can be even more difficult to distinguish genes that are reliably influenced by the condition from spurious genes, since they have lower expression levels.

We contend that NEI scores can help assess genes having minimally changed expression. To demonstrate we focus arbitrarily on the third of the same set of experiments. And we examine low induction genes. In this case we look at genes with log expression ratios between 0.5 and 1.5. In this experimental set there are some 308 genes with expression values in this range. Under these circumstances, it would be difficult to do detailed experimental follow-up of all of these genes. In addition false positives will be abundant.

One way of distinguishing genes that are true positives and false positives is to do replicates. Presumably, the more times a gene achieves expression values greater than 0.5 in other experiments, the more likely it is to be a true positive. Of course, this will not account for systematic error, but should reduce sporadic errors such as the ones caused by aberrations in inducing the experimental condition in that particular trial. In this case we have five other replicates of the same experiment to examine. Since genes with high NEI scores should be true positives, then the experimental replicates for those genes should also have expression values that are greater than 0.5.

In Figure 5.7(a) we have plotted the average number of times the replicated expression values of these low induction genes are greater than 0.5 in other experiments. As the literature-based NEI scores increase, the average number of positive replicates increases as well. Similarly the fraction of genes with at least four out of five positive replicates correlates with the NEI score (Figure 5.7b). This correlation with the reproducibility of a gene's expression suggests that the NEI score can be used to suggest likely true positives, even among genes with low induction.

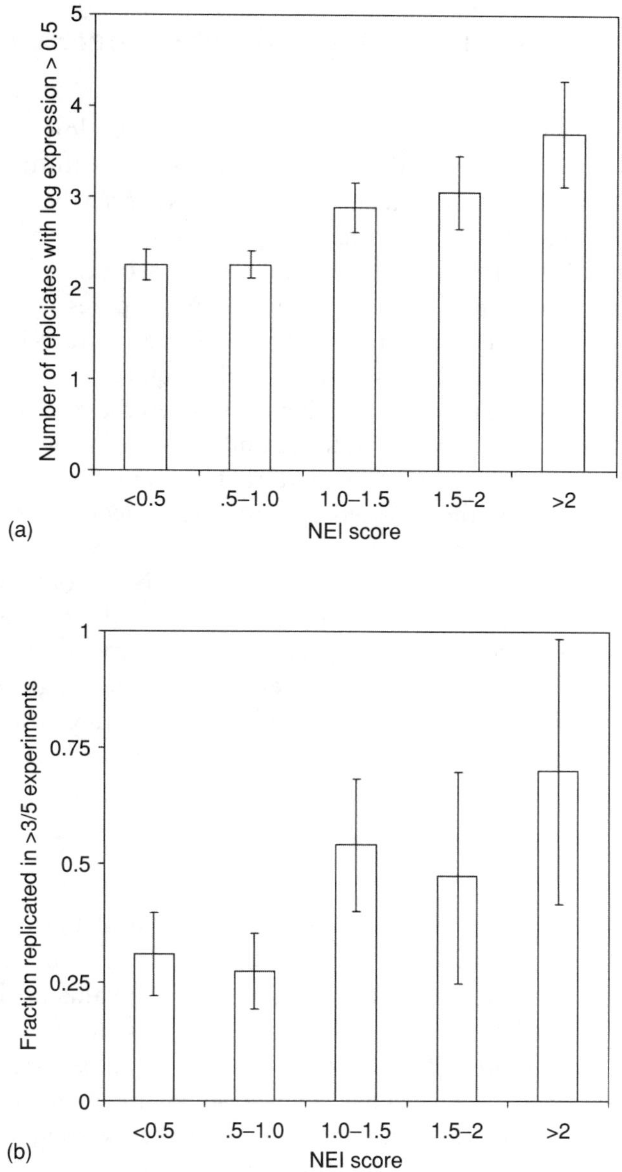

Figure 5.7 *Genes with low induction.* Here we look specifically at genes in experiment 3 that are modestly induced with log expression rations between 0.5 and 1.5 These genes are divided by their NEI scores into five ranges. Genes with high NEI scores are more likely to be true positives. (a) The average number of the five replicates in which those genes have a log expression ratio greater than 0.5. (b) The proportion of genes with a log expression ratio greater than 0.5 in at least four of the five replicates. In both plots 95% confidence intervals are included.

5.9 Assessing experiment quality with literature-based scoring

The single experiment we looked at in Section 5.7 was a lower quality data set than the averaged data of all six. It was apparent that the data from this experiment did not capture a clean physiologic response when we looked at the highest induced genes. They did not appear to represent a clear physiologic process. This was reflected in the lower NEI scores.

Up until now we have been arguing that NEI scores can be used to distinguish false positives and true positives. In addition, NEI scores can be used to compare experiments and assess their quality.

For each of the six experiments we have tabulated the median NEI scores for all genes and for the top 15 induced and repressed genes in Table 5.5. The averaged experiment, which is data that contains the least noise, has the highest scores for the induced and repressed genes, and the lowest median score for all genes. Also for comparison we show the same statistics for a random data set. Not surprisingly, the median NEI score is no different from the median NEI scores of the top 15 induced and repressed genes.

These numbers give us a good sense of the quality of the data. High quality data sets have induced and repressed genes with high median NEI scores. Low quality data sets, on the other hand, may show little enrichment of high NEI scoring genes at the extremes of expression.

5.10 Improvements

Here we have demonstrated a scoring system that evaluates a gene in the context of other genes. To assess the involvement of a gene in a given experiment it looks at the expression of its neighbors. The value of the scientific literature, here, is to help identify what those neighbors are. As we apply more effective strategies to identify functional neighbors, the performance of this method will improve; for example application of latent semantic indexing might improve performance. In addition more effective weighting schemes and better distance metrics will likely improve performance.

In addition, in its current formulation, there is poor accounting for the fact that some articles have hundreds of referring genes while others only have a single one. When constructing a word vector for a gene, there is likely some advantage to down-weighting the influence of articles that refer to many other genes; these articles are likely nonspecific assays. On the other hand, articles that refer to only that gene are likely very valuable and should be up-weighted when creating word vectors. These adjustments to the formulation of word vectors may also increase the performance of the NEI method altogether.

The strategy we are using to identify neighbors requires us to create vectors of words by averaging the content of many different articles. This can be a major disadvantage as the articles may have diverse content. As we average many different articles that talk about different aspects of the genes, valuable signal may be diluted out. A better strategy is to treat the articles separately, and not merge their content into a single vector. We will address this issue more thoroughly in Chapter 6.

Finally, there is the issue that many genes are poorly studied and lack sufficient amounts of literature to create accurate NEI scores for. In these cases, the current formulation we have introduced is less effective. In many cases, these are the most interesting genes to focus on as well. This is one of the real challenges to using any literature-based approach. In Chapter 4 we introduced sequence-based strategies to supply these genes with surrogate references.

5.11 Application to other assays

The framework that we have introduced here is valuable to many genomic assays where the response of thousands of genes to a stimulus is being assayed. In all of these cases, the sheer number of genes examined often requires external corroboration to evaluate whether a positive finding by assay makes sense in the context of the rest of the data. The NEI approach offers a means to do this. For example, the Serial Analysis of Gene Expression (SAGE) assay can also assess expression of genes, but in an alternative fashion (Velculescu, Zhang et al. 1995). The assay results can also be used with NEI scores to assess the reliability of highly expressed genes. Another assay that this approach can be effectively applied to is large yeast-2-hybrid assays where thousands of genes are screened and scored for their ability to bind a single protein (Fields and Song 1989). A more detailed discussion of protein binding is presented in Chapter 10. One would expect that a true positive binding protein should share some biological function with other proteins that are able to bind. NEI scoring can be used to identify proteins that have functionally similar proteins that are also likely binding proteins.

5.12 Assigning keywords that describe the broad biology of the experiment

In this section we introduce a strategy to assign keywords to the gene expression experiment that broadly describe the function of the genes that are either induced or repressed. The goal is to identify words that provide the biologist with a quick sense of the experiment and which genetic processes are active. There are many possible strategies that can effectively accomplish this goal.

In the previous chapters we showed how weighted word vectors for different genes could be averaged together; the greatest valued words in this vector can often be effective keywords for that group. This strategy could be applied in the context of gene expression data if a group of genes could be defined effectively. This is difficult, however, since true positive and false positive genes are mixed together, and it is impossible to create an expression threshold that clearly separates a group of genes that are truly responding to the physiologic condition from non-responsive genes.

Instead we propose another approach in this section. We calculate an expression value for each word. We do this by looking at each article and assigning it an expression value by averaging the genes it refers to. Then the expression value of a word is the average expression value of all of the articles it is in. This is depicted in the schematic in Plate 5.1. The final step is to determine whether or not that expression value for the word is significantly positive or negative.

The average expression of an article i that has references to m_i genes is:

$$a_i = \frac{1}{m_i} \sum g_j$$

where a_i is the averaged expression of the genes j that it refers to, and g_j is the expression of gene j. In matrix form, we define a matrix R (for reference matrix) where the rows correspond to articles, and the columns correspond to genes. If there are N_a articles and N_g genes, then this matrix is $N_a \times N_g$. The entry at any position (i, j) is non-zero only if article i has a reference to gene j. If non-zero entries at each position (i, j) are assigned $1/m_i$, then with this formulation,

$$a_i = \sum r_{ij} \times g_j$$

or

$$A = R \times G$$

where G is a column vector that contains the gene expression values for each of the genes, and A is a column vector that contains the expression values for each of the articles.

Similarly, the expression value of a word i that occurs in m_i articles, each with an expression a_j, can be calculated by averaging the expression of those articles:

$$w_i = \frac{1}{m_i} \sum a_j$$

where w_i is the calculated expression value. In matrix form, we define a matrix T (for text matrix) where the rows are words and the columns are articles. The entry (i,j) is non-zero only if the word is in the article. If the word i is in the article j then the entry (i,j) is set to $1/m_i$. So each entry is the inverse of the number of articles the word appears in. With this formulation:

$$w_i = \sum t_{ij} \times a_j$$

or:

$$W = T \times A$$

where A is a column vector that contains the gene expression values for each of the articles, and W is a column vector that contains the calculated expression values for each of the words. In fact we can see that the expression values for the words W is calculated directly from the gene expression values G:

$$W = T \times R \times G$$

So actually, for each word, its expression is a weighted mean of gene expression values where the weights are contained in the matrix $Y = T \times R$.

Once the mean expression values for each of the words are calculated, the key step is to determine whether the expression values are significant or not. The first step to this is to calculate the variance of the weighted mean for each word:

$$s_i^2 = \sum_j y_{ij}(g_j - w_i)^2$$

Where y_{ij} is the weighted contribution of the gene expression g_j to the mean word expression value, w_i. The next step to calculating the variance of the mean is as follows:

$$\sigma_i^2 = s_i^2 \sum_j y_{ij^2}$$

Once a mean expression value has been calculated for each word, and its mean variance determined, we can determine the number of standard deviations it is away from zero, or its z-score (see Section 2.2):

$$z_i = \frac{w_i - x}{\sqrt{\sigma_i^2}}$$

Here x is the mean expression of all genes, w_i is the mean expression for the word. The square root of the variance is the standard deviation. A word that has an expression value that is 1.97 standard deviations away from zero is significant at the alpha < 0.05 level (assuming a two-tailed test), while a word that is 2.58 standard deviations away from the mean is significant at the alpha < 0.01 level. Since we are looking at on the order of a thousand words, the standard of statistical significance level is much greater. Using the standard Bonferoni correction, we would prefer significance values on the order of 0.05/1000. These values can be achieved with z-scores of 4.05.

Results of the application of this method to the phosphate deprivation data are displayed in Figure 5.8. This strategy provides some clues about the general expression responses of genes involved in different biological processes. There were a total of 1179 words present in more than 300 articles. We selected these common words. Then we used the above equations to calculate the mean expression for each word, and the mean variance. We selected all words with z-scores greater than 4.05 with the greatest mean positive and negative expression. These words are listed in the figure. We have plotted the mean expression value for each word with 99.995% confidence intervals; these confidence intervals include 4.05 mean standard deviations on either side of that value.

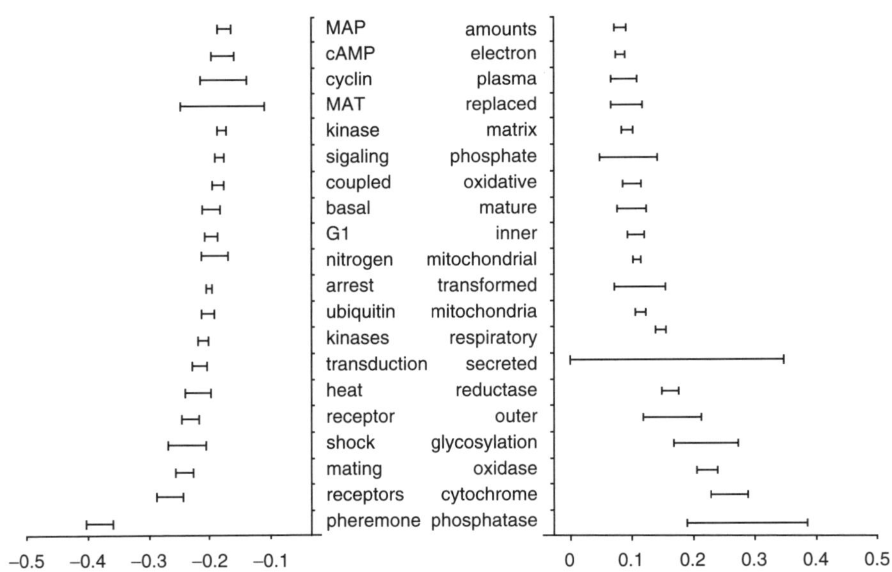

Figure 5.8 *Keywords identified that characterize phosphate deprivation experiments.* Top positively expressed and negatively expressed keywords, plotted with confidence intervals.

The most induced word is "phosphatase". It makes sense that under the condition of phosphate deprivation genes that are phosphatases that free phosphate from other molecules are up-regulated. Not surprisingly the word "phosphate" is also induced, highlighting the critical role this molecule plays in this condition. Many of the other induced words seem to be connected to aerobic respiration and energy generation in the mitochondria. These words include "mitochondria", "respiratory", "electron", "cytochrome" and "oxidase".

On the other hand, the genes that are suppressed are related to mating and cell reproduction. Genes related to communicating and mating with other cells, such as signaling molecules and pheromones, are repressed; the words "MAT", "pheromone", "signaling", and "receptors" are down-regulated. Also there is a suggestion that cell cycle associated genes are also repressed during starvation; words like "cyclins" and "G1" have negative log expression ratios. During a starvation state cellular replication and reproduction is low priority. Finally kinase proteins, responsible for phosphorylation of proteins, is down-regulated when phosphate is scarce. These keywords seem to offer some insight as to what is happening in this condition.

References

Ball, C. A., I. A. Awad, et al. (2005). "The Stanford Microarray Database accommodates additional microarray platforms and data formats." *Nucleic Acids Res.* 33 **Database Issue**: D580–2.

Fields, S. and O. Song (1989). "A novel genetic system to detect protein-protein interactions." *Nature* **340**(6230): 245–6.

Lee, M. L., F. C. Kuo, et al. (2000). "Importance of replication in microarray gene expression studies: statistical methods and evidence from repetitive cDNA hybridizations." *Proc. Natl. Acad. Sci. USA.* **97**(18): 9834–9.

Novak, J. P., R. Sladek, et al. (2002). "Characterization of variability in large-scale gene expression data: implications for study design." *Genomics* **79**(1): 104–13.

Saldanha, A. J., M. J. Brauer, et al. (2004). "Nutritional homeostasis in batch and steady-state culture of yeast." *Mol. Biol. Cell.* **15**(9): 4089–104.

Tu, Y., G. Stolovitzky, et al. (2002). "Quantitative noise analysis for gene expression microarray experiments." *Proc. Natl. Acad. Sci. USA.* **99**(22): 14031–6.

Velculescu, V. E., L. Zhang, et al. (1995). "Serial analysis of gene expression." *Science* **270**(5235): 484–7.

6 Analyzing groups of genes

The analysis of large-scale genomic data (such as sequences or expression patterns) frequently involves grouping genes based on common experimental features. The goal of manual or automated analysis of genomics data is to define groups of genes that have shared features within the data, and also have a common biological basis that can account for those commonalities. In utilizing algorithms that define groups of genes based on patterns in data it is critical to be able to assess whether the groups also share a common biological function. In practice, this goal is met by relying on biologists with an extensive understanding of diverse genes that decipher the biology accounting for genes with correlated patterns. They identify the relevant functions that account for experimental results. For example, experts routinely scan large numbers of gene expression clusters to see if any of the clusters are explained by a known biological function. Efficient definition and interpretation of these groups of genes is challenging because the number and diversity of genes exceed the ability of any single investigator to master. Here, we argue that computational methods can utilize the scientific literature to effectively assess groups of genes. Such methods can then be used to analyze groups of genes created by other bioinformatics algorithms, or actually assist in the definition of gene groups.

In this chapter we explore statistical scoring methods that score the "coherence" of a gene group using only the scientific literature about the genes—that is whether or not a common function is shared between the genes in the group. We propose and evaluate such a method, and compare it to some other possible methods. In the subsequent chapter, we apply these concepts to gene expression analysis.

The major concepts of this chapter are described in the frame box. We begin by introducing the concept of functional coherence. We describe four different strategies to assess the functional coherence of a group of genes. The final part of the chapter emphasizes the most effective of these methods, the neighbor divergence per gene. We present a discussion of its performance properties in general and on its robustness given imperfect groups. Finally we present an example of an application to gene expression array data.

1) Functional coherence of a gene group	a) Performance
2) Word distribution divergence	b) Score robustness to imperfect groups
3) Best article score	c) Application to expression data
4) Neighbor divergence	
5) Neighbor divergence per gene	

6.1 Functional coherence of a group of genes

The main challenge that we address in this chapter is creating a computational method that analyzes scientific literature about a group of genes to determine whether the group is a biologically meaningful one. The goal is to create a method that can quickly assess the biological significance of a group of genes based on scientific text. We have many statistical methods available to us, some of which were discussed in Chapter 2, that can help us assess and create group of genes with statistical similarity in experimental data. But statistical significance and biological significance do not always correspond. Methods such as those we introduce here help to assess the biological significance of a group of genes based on the scientific literature. These methods can be used after the application of statistical algorithms to select the groups that are biologically meaningful that also correlate with patterns in the data. Alternatively, experimental data analysis can be used in conjunction with literature-based approaches to create more meaningful groups—that is, groups can be defined with optimal biological coherence as well as similarity in the experimental data. We can achieve these goals by using two measures of similarity: one based on the experimental data and a second on the biological literature, and optimizing both of them.

In this chapter we will use the term "functional coherence" to describe the degree to which a set of genes have a common biological basis. In Table 6.1 we have listed a set of functionally coherent genes. These are all of the genes in yeast that are the *DNA-dependent ATPase* genes in yeast; we obtained these genes from Gene Ontology (Ashburner, Ball et al. 2000). These genes convert ATP to ADP and generate energy that is then used to manipulate DNA. Since these genes share this function in common, we would say this group is a functionally coherent group.

Each gene is listed alongside a critical article reference that suggests that function for the gene. The shared function of these genes is immediately apparent by glancing at these references. We can see that many of the same

Table 6.1 *DNA dependent ATPase genes in yeast.* This table lists the genes with the DNA dependent ATPase function in yeast as indicated in the Gene Ontology database. We list the key reference cited in Gene Ontology that suggests that biological function in the second column. In the final column we list the total number of references that each gene has.

Gene	Article	Article references
RAD16/YBR114W	**Guzder SN, et al. (1998)** The DNA-dependent ATPase activity of yeast nucleotide excision repair factor 4 and its role in DNA damage recognition. *J Biol Chem* **273**(11):6292–6	66
RAD18/YCR066W	**Bailly V, et al. (1997)** Yeast DNA repair proteins Rad6 and Rad18 form a heterodimer that has ubiquitin conjugating, DNA binding, and ATP hydrolytic activities. *J Biol Chem* **272**(37):23360–5	96
RAD26/YJR035W	**van Gool AJ, et al. (1994)** RAD26, the functional S. cerevisiae homolog of the Cockayne syndrome B gene ERCC6. *EMBO J* **13**(22):5361–9	34
RAD54/YGL163C	**Petukhova G, et al. (1999)** Yeast Rad54 promotes Rad51-dependent homologous DNA pairing via ATP hydrolysis-driven change in DNA double helix conformation. *J Biol Chem* **274**(41):29453–62	189
RAD7/YJR052W	**Guzder SN, et al. (1998)** The DNA-dependent ATPase activity of yeast nucleotide excision repair factor 4 and its role in DNA damage recognition. *J Biol Chem* **273**(11):6292–6	67
RDH54/YBR073W	**Petukhova G, et al. (2000)** Promotion of Rad51-dependent D-loop formation by yeast recombination factor Rdh54/Tid1. *Genes Dev* **14**(17):2206–15	26
RIS1/YOR191W	**Zhang Z and Buchman AR (1997)** Identification of a member of a DNA-dependent ATPase family that causes interference with silencing. *Mol Cell Biol* **17**(9):5461–72	3

words appear in the titles of all of these genes. Ideally, functionally coherent gene groups have two properties: (1) all of the genes have the same function, and (2) all of the genes with that function are contained in the group itself. In practice, text-based functional scoring schemes can only approximate this ideal. For example, a group of genes similar to the one depicted in Table 6.1 that contains two additional unrelated genes may receive a lower coherence score, but likely still a significant one. In the same way a group of genes similar to the one in the table but missing two of the genes will also receive a significant score. For our purposes, this is actually a helpful property as experimental methods rarely generate perfect groups of genes and more often than not we are attempting to detect approximate groups. However, we would prefer that, as a group becomes closer to ideal, its functional coherence score does improve.

Another caveat about functionally coherent groups of genes is that they may be broad or very narrow in scope. For example, a group of genes containing all of the genes involved in metabolism can be a large, but coherent group of genes. On the other hand, a smaller subset of genes containing only carbohydrate metabolism is still functionally coherent, even though it is a much smaller set of genes. Going further, the even smaller subset of genes involved in glycolysis is also equally functionally coherent. So this can be confusing since all metabolism genes may include thousands of genes, whereas all glycolysis genes may include ten genes or so. In the context of the analysis of experimental data, both large broad groups and smaller narrow functional groups are equally important, as either type can be affected by different stimuli and both are important to recognize.

Recognizing coherent gene groups from the literature is a challenging problem, since there are disparities in the literature about genes. A given gene may have many relevant documents or none, and the documents about it may cover a wide spectrum of functions. This issue is addressed in great detail in Chapter 1. Some genes have been extensively studied while others have only been recently discovered, and may not have any available articles. This disparity is apparent even in the group listed in Table 6.1, which is an unusually well studied group as evidenced by the number of article references available for each gene. Note, however, that while most genes have 20–100 references, one has 189 and another has only three.

Additionally, most genes have multiple functions, and this is reflected in the literature. References about a given gene might include articles that discuss the gene's sequence, phenotype, molecular function, location in the cell, one or more of its biological functions, or the structure of its protein product. For example, consider the gene RDH54 that has 26 relevant articles, the most recent of which are listed in Table 6.2. Of these articles none of them address the *DNA dependent ATPase* function of the gene. The articles address a variety of different functions including meiosis, recombination, DNA double strand break repair, and a sequencing paper. These other articles are not relevant to this issue. So effectively understanding the relationship of the genes in Table 6.1 relies on our ability to recognize and focus on those key articles that tie these genes together. So in this chapter, unlike the previous ones, we treat each article as an independent source of

Table 6.2 *Recent articles about RDH54.* Here we have listed 16 of the most recent articles about this gene. These articles speak to the diverse and complex functionality of this gene, and none of these articles addresses the genes DNA-dependent ATPase function in an obvious manner.

Shinohara M, et al. (2003) Crossover interference in Saccharomyces cerevisiae requires a TID1/RDH54- and DMC1-dependent pathway. *Genetics* 163(4):1273–86

Lee SE, et al. (2003) Yeast Rad52 and Rad51 recombination proteins define a second pathway of DNA damage assessment in response to a single double-strand break. *Mol Cell Biol* 23(23):8913–23

Kellis M, et al. (2003) Sequencing and comparison of yeast species to identify genes and regulatory elements. *Nature* 423(6937):241–54

Fukuda T, et al. (2003) VDE-initiated intein homing in Saccharomyces cerevisiae proceeds in a meiotic recombination-like manner. *Genes Cells* 8(7):587–602

Catlett MG and Forsburg SL (2003) Schizosaccharomyces pombe Rdh54 (TID1) acts with Rhp54 (RAD54) to repair meiotic double-strand breaks. *Mol Biol Cell* 14(11):4707–20

Symington LS (2002) Role of RAD52 epistasis group genes in homologous recombination and double-strand break repair. *Microbiol Mol Biol Rev* 66(4):630–70, table of contents

Shor E, et al. (2002) Mutations in homologous recombination genes rescue top3 slow growth in Saccharomyces cerevisiae. *Genetics* 162(2):647–62

Miyagawa K, et al. (2002) A role for RAD54B in homologous recombination in human cells. *EMBO J* 21(1–2):175–80

Signon L, et al. (2001) Genetic requirements for RAD51- and RAD54-independent break-induced replication repair of a chromosomal double-strand break. *Mol Cell Biol* 21(6):2048–56

Lee SE, et al. (2001) The Saccharomyces recombination protein Tid1p is required for adaptation from G2/M arrest induced by a double-strand break. *Curr Biol* 11(13):1053–7

Klein HL (2001) Mutations in recombinational repair and in checkpoint control genes suppress the lethal combination of srs2Delta with other DNA repair genes in Saccharomyces cerevisiae. *Genetics* 157(2):557–65

Sung P, et al. (2000) Recombination factors of Saccharomyces cerevisiae. *Mutat Res* 451(1–2):257–75

Shinohara M, et al. (2000) Tid1/Rdh54 promotes colocalization of rad51 and dmc1 during meiotic recombination. *Proc Natl Acad Sci U S A* 97(20):10814–9

Petukhova G, et al. (2000) Promotion of Rad51-dependent D-loop formation by yeast recombination factor Rdh54/Tid1. *Genes Dev* 14(17):2206–15

Haber JE (2000) Recombination: a frank view of exchanges and vice versa. *Curr Opin Cell Biol* 12(3):286–92

Haber JE (2000) Partners and pathwaysrepairing a double-strand break. *Trends Genet* 16(6):259–64

information. Combining the signal from all of the articles about a gene into a single word vector dilutes the important signal that we are interested in.

The best performing method is neighbor divergence per gene (NDPG). It requires only a corpus of documents relevant to the genes being studied (e.g. all genes in an organism) and an index connecting the documents to appropriate genes. Given a group of genes, NDPG assigns a numerical score indicating how "functionally coherent" the gene group is from the perspective of the published literature (Raychaudhuri, Schütze et al. 2003). The method was tested by assessing its ability to distinguish 19 known functional gene groups from 1900 randomly assembled groups. NDPG achieves 95% sensitivity at 100% specificity, comparing favorably to other tested methods.

An alternative approach to assessing the functional coherence of a gene group is to cross-reference it against predefined groups of related genes that have been compiled automatically from the literature or by manual annotation. The Gene Ontology (GO) consortium and the Munich Information Center for Protein Sequences (MIPS) provide vocabularies of function and assign genes from multiple organisms the relevant terms (Ashburner, Ball et al. 2000; Mewes, Frishman et al. 2000). Genes that are assigned the same term constitute a functional group of genes. However, such resources may not be comprehensive and up to date at any given time, and it is also laborious to maintain the vocabulary and the gene assignments. The literature-based method introduced here requires only a set of references associated with genes. It requires no precompiled lexicons of biological function, previous annotations, or co-occurrence in the literature. It is kept current and up to date if it is provided a current literature base. Furthermore, this method can be applied to any arbitrary set of genes, as long as an index of gene-article associations is provided. These precompiled sets of genes are very helpful, however, in that they provide a nice set of functionally coherent groups that can be used as a gold standard to evaluate the literature methods that we propose.

6.2 Overview of computational approach

The neighbor divergence per gene (NDPG) and the other methods that we test in this section require only a corpus of articles relevant to the studied genes (e.g. all genes in an organism) and a reference index connecting the articles to appropriate genes. Such reference indices are often available online from genome centers (see Chapter 1) or can be compiled automatically by scanning titles and abstracts of articles for gene names (Jenssen, Laegreid et al. 2001). Given a group of genes, these methods assign a numerical score indicating how "functionally coherent" the gene group is from the perspective of the published literature.

The intuition behind NDPG involves recognizing key articles that are about the function represented in the group. Suppose a group of genes shares some specific function, such as *DNA-dependent ATPase*, and contains all of the genes with that function. An article germane to that function must refer to at least one of the genes in the group. Furthermore, other articles that are semantically similar will pertain to the same function and will also refer to genes in the same group. In Figure 6.1 we have displayed a schematic group and its articles. The articles represented by the darkened boxes are the key articles that tie these genes together. Presumably there is similarity between these articles. These articles are not unlike the articles listed in Table 6.1 that are the key articles for that group. The other articles represented by the light boxes are like the articles listed in Table 6.2; these articles address the other facets of the genes.

NDPG assigns a functional coherence score to a group of genes based on the literature. It uses document distance metrics to calculate semantic neighbors; two articles are semantic neighbors if there is similar word usage in each of them. First, semantic neighbors are pre-computed for each article in the corpus. Given a gene group, each article's relevance to the group

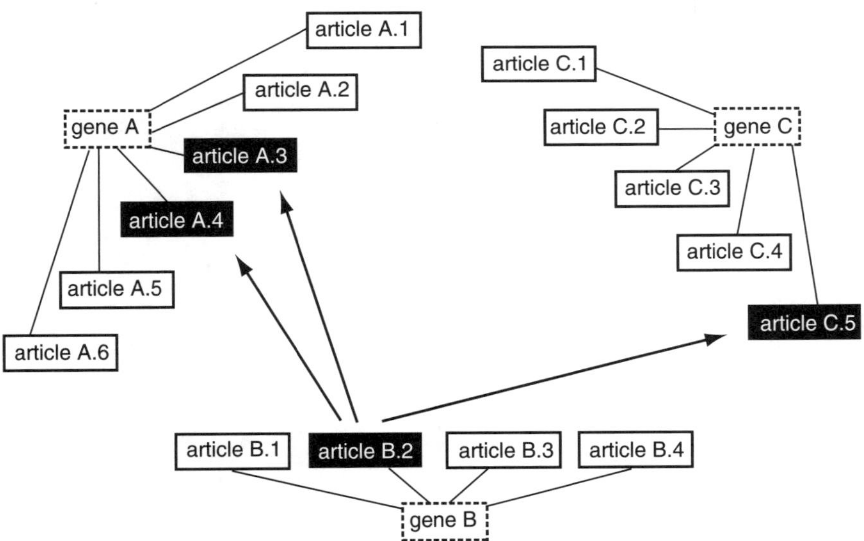

Figure 6.1 *Finding the key articles that link a gene group together.* This is graphical depiction of a small gene group of three genes with the function *DNA-dependent ATPase* (boxes with dotted boundaries). The genes are connected to their respective article references (boxes with solid boundaries). Articles that talk specifically about the *DNA-dependent ATPase* genes are represented as dark boxes with white lettering. For all genes, only a few of the referenced articles are pertinent to this aspect of the gene. The arrows are used to indicate the semantic neighbors of "article B.2", a *DNA-dependent ATPase* article. The significance of this article to the group's unifying function becomes apparent when we notice that many of its semantic neighbors, other articles about the same function, refer to other genes in the same group.

is scored by counting the number of its neighbors that have references to genes in the group. If the group represents a coherent biological function, the articles that discuss that function will have many referring neighbors within the group and therefore score high (see Figure 6.2). Other articles that address biological functions that are unrelated to the group function will score low. If a few of the articles referring to a gene are high scoring articles, then the gene has a function that is relevant to that of the group. For each gene in the group, NDPG scores its functional relevance to the group by comparing article scores of its references to an expected random distribution of article scores; the difference between the two distributions is quantified with the KL – divergence measure. The NDPG measure of functional coherence of a gene group is the mean divergence of all of the genes in the group.

The key aspect of NDPG that makes it very effective is that it is an article-based approach; its success hinges on the presence of a few key articles that

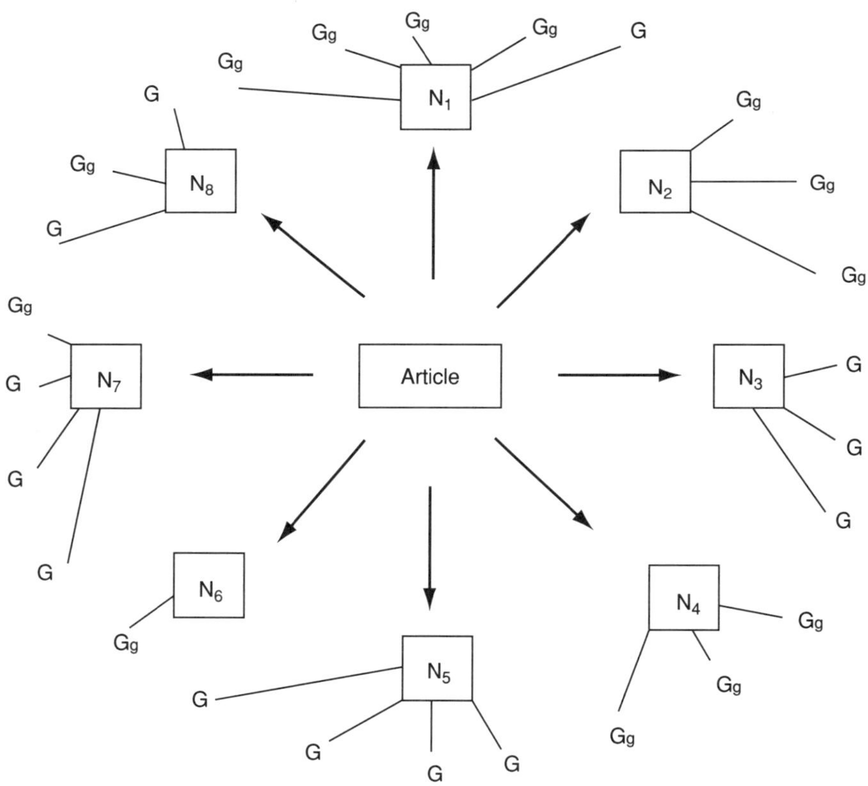

Figure 6.2 *Scoring an article's semantic content for relevance to a gene group.* For each article we look at its semantic neighbors. Here we score the central article; it has eight semantic neighbors indicated by the boxes with N_i labels. In principle the score of an article is the number of its neighbors that refer back to genes in the original group. Here, we indicate group genes with G_g and non-group genes with G. In practice, neighbor articles might refer to multiple genes, some of which are in the group and others of which are not.

unifies the group of genes given. This addresses the issue of recognizing and avoiding irrelevant articles in scoring groups of genes. A second key aspect of NDPG is that the score is a mean divergence of the scores of each of the genes; each of the genes, regardless of whether they have a few or many articles, make an equal contribution to the mean. This addresses the issue of the biases in the amount of literature present per article.

We compare this method to other seemingly promising approaches that lack these key features. We demonstrate a method that is based on the distribution of words. In this method articles are not treated individually. Rather, all of the words from all of the articles are combined together, and a distribution of words among these articles is defined. The theory is that if the genes in this group define a coherent function, the distribution of words describing these genes will be considerably different from the baseline distribution of words in biology.

A second strategy is to look at the single key article that is most relevant to a group of genes. The idea is that a relevant article will have article neighbors that refer to genes in the group of genes. Unlike the prior strategy, this is an article-based strategy. But it lacks robustness as it looks only at a single article.

A third strategy is to score all of the articles in the corpus for its relevance to the group. A significant group that is biologically meaningful should have a larger than randomly expected number of articles with high scores.

6.3 Strategy to evaluate different algorithms

NDPG calculation of a gene group requires a corpus of documents relevant to all genes in the organism, and a reference index indicating the articles that are germane to each gene. As in the previous chapters, the documents are PubMed abstracts. The title and abstract fields in the PubMed records are the only ones used. Those words that are present in more than four abstracts and fewer than 10,000 abstracts are considered as vocabulary words. Abstracts are converted into vectors of word counts. All experiments described below were conducted in *S. cerevisiae*, or baker's yeast. For this study we used a reference list that contained PubMed abstract references to yeast genes (Cherry, Adler et al. 1998). The reference list included 20,101 articles with 50,860 references to 4205 genes.

To evaluate NDPG and compare it to the other approaches described in the previous section, we used 19 known functional yeast gene groups. We also devised 1900 decoy random yeast gene groups. We compared the four methods by scoring all groups. A good method should assign high scores to functional groups and lower scores to random groups.

The functional groups were devised by selecting 19 Gene Ontology terms relevant to yeast biology; all terms were biological process terms. Since GO is a hierarchical vocabulary, each group was defined as the set of genes assigned either the listed term or more specific descendants of that term. The GO terms and some properties of the groups they correspond to are described in Table 6.3(a). The gene groups selected for this study varied vastly in number and content. This diversity is representative of gene groups that experimental procedures may derive. The number of genes per group range from 6 to over 600. Incidentally, many of the genes were members of more than a single functional gene group (Table 6.3b). This underscores the multiple functionality that many genes have.

We assembled a set of 1900 random gene groups as decoy gene groups. For each functional gene group, 100 gene groups were created with the

Table 6.3 *A description of the functional gene groups used to evaluate functional coherence methods.* (a) Each of the gene groups was devised from a Gene Ontology code. In the first column we list the function of the associated code. In the second column we list the number of yeast genes with that function. In the final column we list the number of articles referenced in total by those genes with that function. (b) Many of the genes are in more than one of the groups in (a). This table lists the number of genes in multiple functional groups. Many genes are not in a single functional group. Some genes are in as many as six functional groups.

(a)

Functional classification	Genes	Unique articles referenced
signal transduction	94	1944
cell adhesion	6	59
autophagy	16	55
budding	74	979
cell cycle	341	4438
biogenesis	459	3840
shape size control	54	1014
cell fusion	89	1470
ion homeostasis	43	363
membrane fusion	6	209
sporulation	27	553
stress response	94	1866
transport	313	2708
amino acid metabolism	78	1221
carbohydrate metabolism	90	1855
electron transport	8	187
lipid metabolism	90	715
nitrogen metabolism	15	229
nucleic acid metabolism	676	6674

(b)

Number of groups/gene	0	1	2	3	4	5	6
Number of genes	2412	1242	386	113	40	9	3

same number of random genes. This was done to insure that decoy groups varied as dramatically in size as the functional groups.

In this study we evaluated several different methods to score the functional coherence of a gene group. These methods are described in detail below. Each method was used to score the 1900 decoy gene groups and the 19 functional gene groups. A good method assigns a score to each functional group that exceeds the score of the other 1900 groups. For comparison, we calculated the precision and recall of a method at different score cutoff levels. The precision is the number of functional groups scoring above the cutoff divided by the number of total groups scoring above the cutoff. The recall is the number of functional groups scoring above the cutoff divided by the total number of functional groups.

6.4 Word distribution divergence

In Chapter 2 we defined word vectors for genes and groups of genes by averaging word vectors for many different articles together. These amalgamated vectors were simple to use and effective in sequence analysis and in expression analysis. For example, in the previous chapter we analyzed single condition gene expression data by creating word vectors for each gene that were averages of the word vectors for referring articles. The crux of this approach was that we approached the literature by merging a large collection of pertinent literature together. These approaches are conceptually simple and easy to implement.

The disadvantage of these approaches is that merging articles eliminates their uniqueness. Merging together such text information often gives the impression of creating more robust word vectors. But in practice the content of the articles is diverse, even those articles describing the same gene. As a result combining the information from many articles into a single representation for a gene can actually dilute the signal. For example, a very nonspecific collection of words may result from merging articles about the protein structure of a gene product, the human diseases that a gene has been implicated in, and a broad screen that had been used to identify the association of that gene with others.

As a starting point to address automatic functional coherence assessment, we describe the word distribution divergence (WDD) method (Raychaudhuri, Schutze et al. 2002). The WDD strategy is similar to those strategies in prior chapters, where word counts for many articles for many genes are simply merged together to create a single representation for the gene group. This method requires calculation of and comparison of two distributions of words. The first, f, is a distribution of words used in all articles referring to genes in the group; the second, g, is a distribution of words referring to genes outside the same group. The thinking behind this approach is that if a specific function is represented in the group then words corresponding to that function will be enriched among the articles referring to genes in that group. The other distribution of articles referring to genes outside the group will have a paucity of those same words. Consequently there will be differences between these two distributions that become increasingly dramatic as the gene group becomes more functionally coherent. A group of ribosomal genes, for example, should have referring articles with words like "ribsome" and "tRNA" occurring considerably more frequently than in the other articles.

The distribution f is computed from words in abstracts referring to genes within the group; counts of each word are divided by the total number of words in these abstracts:

$$f(w_i) = \frac{\sum_j w_{ij}}{\sum_j n_j}$$

where w_{ij} is the number of times that word i appears in document j, and n_j is the total number of words in document j. In practice, articles may refer to multiple genes, some of which are contained within the group and others outside. To account for this we calculate the parameter fr for each article, where:

$$fr_{k,\,g} = \frac{r_{k,\,g}}{r_k}$$

where $r_{k,\,g}$ is the number of genes in the gene group g that the document k refers to, r_k is the number of genes that document k refers to altogether; $fr_{k,\,g}$ is the fractional reference for document k to group g. In practice, to correctly account for articles having gene references both inside and outside a given group, we calculate the empirical probability distribution f by weighting word occurrences:

$$f(w_i) = \frac{\sum_j fr_{j,\,g} w_{ij}}{\sum_j fr_{j,\,g} n_j}$$

A distribution g is computed similarly for all abstracts referring to other genes outside the group:

$$g(w_i) = \frac{\sum_j (1 - fr_{j,\,g}) w_{ij}}{\sum_j (1 - fr_{j,\,g}) n_j}$$

To avoid extremely low probability words that could inadvertently bias the outcome both distributions are smoothed. This is particularly an issue if the gene group has only a small number of relevant articles, and certain words are not seen in those articles spuriously. Under those circumstances the empirical distribution will have zero probability. Both distributions are smoothed with Dirichlet priors assuming 300 prior words distributed according to a baseline distribution, *b*:

$$f_s(w_i) = \frac{f(w_i)\left(\sum_j fr_{j,\,g} n_j\right) + b(w_i) \times 300}{\left(\sum_j fr_{j,\,g} n_j\right) + 300}$$

The baseline distribution of each word is computed by dividing its count in all abstracts by the total count of all words in all abstracts. The KL-divergence between these two distributions of words is then computed to quantify the difference between them; the gene group distribution f is treated as the observed distribution. The divergence value is used as a measure of functional coherence:

$$D(f_s \parallel g_s) = \sum_i f_s(w_i) \log_2 \left[\frac{f_s(w_i)}{g_s(w_i)}\right]$$

Therefore, WDD is an information theoretic measure of the disparity between the two word distributions. If a subset of rare words is used significantly more inside the group than outside the group, then these words may be indicative of some biological function within the gene group. Therefore WDD should be sensitive to the presence of biological function in the gene group.

Figure 6.3 plots the precision and recall at different cutoff levels for WDD and other methods for comparison. As the cutoff score is selected to be more stringent, some functional groups are not obtained and therefore recall is lower. But, most random groups fail to make the cutoff and the precision is higher. WDD only achieves 10.5% recall (2 out of 19 functional groups) at 8.3% precision on the same data set; this is equivalent to 10.5% sensitivity at 98.9% specificity. The method does detect some signal since it achieves 63% sensitivity at 64% specificity; a totally random method would be 50% sensitive at 50% specificity.

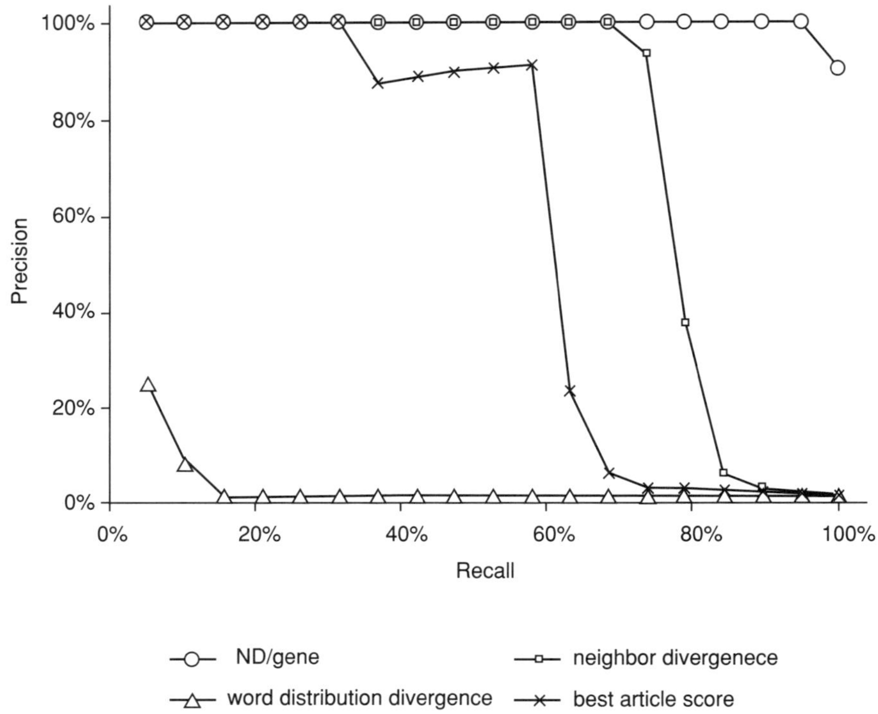

Figure 6.3 *Precision–recall plot for each of the functional coherence scoring methods.* We used each method to score the functional coherence of the 19 functional gene groups and the 1900 random gene groups. We calculated and plotted precision and recall at cutoff scores of different stringency. There is a tradeoff between precision and recall. More stringent cutoff values select fewer true functional groups and recall (or sensitivity) is compromised; however less stringent cutoff values cause many random groups to be selected inappropriately and precision is compromised. An ideal precision–recall plot achieves 100% precision for every value of recall. The "neighbor-divergence" method is closest to the optimal curve.

This method performs relatively poorly. While an individual article may address a single aspect of a gene's function, different articles referring to the same gene may discuss many different biological functions. Consequently, pooling all of the articles referring to a gene results in an uninformative distribution of words. If all articles written about a gene addressed the same function, this method would have been more successful.

6.5 Best article score

To demonstrate the value of article-based approaches, we introduce one simple approach. This approach shows a dramatic improvement from the word-based approach described in the last section. The best article score (BAS) works by scoring the semantics of each article in the corpus for relevance to the group of genes and then using the score of the best one as a coherence measure. The score of an article is calculated by first identifying its semantic neighbors, and then counting the number of neighbors that refer to the genes in the group. The highest article score is used as a measure of a gene group's functional coherence. Here, we will identify 199 neighbors for each article, and in this case scores range from 0 to 199. While this method is relatively simple compared to NDPG, testing its performance demonstrates a definitive improvement between word-based and article-based approaches to this same problem. It is limited, however, in that it utilizes only the score of a single article rather than the distribution of article scores.

For each document, the 199 most similar documents (not including the article itself) are pre-computed. To quantify the similarity between two documents we use the cosine between the two weighted document word vectors. To reduce the impact of common words, word vectors are first converted into inverse document frequency weighted word vectors as described in Chapter 3:

$$w_{i,j} = \begin{cases} (1 + \log_2(tf_{i,j}))\log_2(N/df_i) & \text{if } tf_{i,j} > 0 \\ 0 & \text{if } tf_{i,j} = 0 \end{cases}$$

where $w_{i,j}$ is the weighted count of word i in document j, $tf_{i,j}$ is the term frequency of word i in document j, df_i is the document frequency of word i, and N is the total number of documents. Document similarity is the cosine of the angle between two weighted article vectors.

Given a gene group, we assign a score, S_i, to each document i. The score is the count of semantic neighbors that refer to group genes. The higher the score of an article, the more relevant is its content to the shared biological function that is represented in the group of genes. Coherent groups of genes will cause a small set of germane articles to have extremely high scores. Calculation of article scores is displayed schematically in Figure 6.2.

Practically, semantic neighbor documents in the data set may refer to multiple genes rather than a single one. In many circumstances only some of those genes may be contained in the group. If all of those genes are contained in the gene group, then it makes sense to include the neighbor article in the count of referring neighbors. If none of them are in the group, then it is not included in the count at all. If only some of the genes are contained, then the neighboring document is included as a fraction:

$$fr_{k,g} = \frac{r_{k,g}}{r_k}$$

where $fr_{k,\,g}$ is the fractional reference for document k to group g.

To obtain the document score, the referring fractions of the 199 neighbors are summed and rounded to the nearest integer.

$$S_{i,\,g} = \mathrm{round}\left(\sum_{j=1}^{199} fr_{\mathrm{sem}_{i,\,j},\,g}\right)$$

where $S_{i,\,g}$ is the score for document i for a group g calculated by rounding and summing the fractional reference of its 199 neighbor documents whose indices are $\mathrm{sem}_{i,\,j}$. The value $S_{i,\,g}$ is an integer that ranges between 0 and 199.

In practice, in defining the neighbors it is important to avoid articles that refer to the same genes. Certain genes have large numbers of article references; often many of these articles can be very similar to each other as they may address the same issues about the same gene and on many occasions can even be written by the same authors. In scoring articles for relevance to a group of genes, neighbor articles that share reference to the same gene can confound the analysis. Any group containing that particular gene will cause those article to receive artificially high scores. To avoid this issue the method uses a simple filter in the selection of the 199 neighbors for each document. The method selects neighbors of a given document only from those other documents that refer to at least one additional gene not in the set of genes referred to in the given document.

The best article score (BAS) is similar to NDPG in that articles are scored for relevance against the gene group by counting the number of referring semantic neighbors. In BAS, however, only the single highest article score is used as a score for the group. This method outperforms WDD, since it does not combine a signal from many different articles, but rather considers the articles individually. BAS achieves 31.6% recall at 100% precision (100% specificity) (Figure 6.3). This is a dramatic improvement.

This method searches for articles that have semantic content that is relevant to the group. The advantage of this approach is that articles are treated as individuals. This approach is more appropriate for the problem, since genes are often multifaceted. Scientific articles, however, tend to be focused on the subject they are addressing, and appropriate semantic neighbors can easily be found using good weighting schemes and distance metrics. The BAS method is limited since it uses the score of only a single article; this ignores other high scoring articles that should be abundant if the gene group truly represents a function. The NDPG method relies on the referring neighbor principle also, but in contrast obtains greater statistical power by considering the scores of many articles and not just the extreme valued ones.

6.6 Neighbor divergence

In neighbor divergence (ND) all the documents are scored as described above. (A small technical difference is that 200 neighbors per article are used, and one of the neighbors is the seed article itself.) Like the BAS approach, this too is an article-based approach, but considers the score of all articles instead of just the best scoring article. The empirical distribution of article scores is obtained and examined. The idea is that if there is functional coherence, there should be an excess of high scoring articles. The number of these high scoring articles should exceed what we might expect by random. The random distribution of article scores is modeled with the Poisson distribution. The KL divergence between the entire empirical distribution of all of the document scores and the theoretical Poisson distribution is calculated and reported as a measure of functional coherence.

6.6.1 Calculating a theoretical distribution of scores

In this subsection we estimate the expected distribution of article scores if the gene group has no coherent functional structure. This gives us a basis of comparison for the article scores we obtain. If the empirically obtained article scores for a given set of genes are similar to what we would expect for a random, incoherent group of genes, then we might assume that the given group of genes is incoherent.

To calculate the theoretical distribution of article scores for a random group of genes, we begin by assuming that the chance that an article will refer to a gene in a group is a purely independent random statistical event with a probability q. We can estimate the probability q empirically as the number of articles referring to the group divided by the total number of articles. In practice, since many articles refer to multiple genes:

$$q = \frac{\sum_{i=0}^{N} fr_{i,g}}{N}$$

where fr is the fraction of genes per article that refer to genes in the group.

Now consider that there are 200 neighbor articles, and the article score is the number of articles that have references to genes in the group. If the group is completely random, then this is akin to determining the number of successful binary trials each with probability q of success. Since the trials are independent, the probability distribution can be given with the binomial distribution. The mean score of an article should be $\lambda = 200 \times q$. Under most circumstances the group of genes is a small number compared to the number of genes in the data set, and q is relatively small. A Poisson distribution would estimate this distribution accurately for small values of q (see Section 2.2). In this case:

$$P(S = n) = \frac{\lambda^n}{n!} e^{-\lambda}$$

The Poisson distribution gives us a distribution of expected article scores if the group of genes was totally random.

6.6.2 Quantifying the difference between the empirical score distribution and the theoretical one

After all of the articles are scored, an empirical distribution of article scores is tabulated. If the group contains no functional coherence, the distribution of scores should be similar to the Poisson distribution. The neighbor divergence (ND) functional coherence of the group is the KL divergence between the empirical distribution of article scores and the Poisson distribution. Here the empirical distribution is the observed distribution, and the Poisson distribution is the theoretical one. As the number of high scoring articles increases with the coherence of the functional group, the empirical distribution of article scores appears less and less like the random distribution.

On the same data set ND achieves 68% recall at 100% precision. This is an improvement from the best article score strategy. This strategy has the advantage that it is an article-based approach. Unlike the BAS method, it takes advantage of all article scores. However, the disadvantage is that it treats all articles equally in the score. Consequently genes with many articles will weigh heavily on the score, while genes lacking significant numbers of articles will have minimal influence on the score. So the ND score can be influenced by well-studied genes. In addition, since it is an article distribution based approach, larger groups with more articles may have larger significance scores. This is not an ideal property—since a small but very coherent group of genes might be considered as coherent as a large but broad group.

6.7 Neighbor divergence per gene

Neighbor divergence per gene (NDPG) is very similar to neighbor divergence. To calculate the NDPG score we begin by obtaining the article scores for each gene in the group. For *each gene*, we calculate a distribution of article scores. Then we calculate the divergence between the distribution of article scores for each gene in the group and the Poisson distribution calculated in the previous section. The divergence of article scores for each gene is an independent measure of that gene's relevance to the function of the group. The NDPG score is the average divergence for each of the genes. Each gene in the group, should it be relevant to the dominant group function, should have referring documents that score high.

The idea is that we are looking for significant articles among the references for each gene. If there are significant articles, the divergence of the article scores for a gene should be a larger number. If many of the genes have articles that are high scoring and relevant to the group's function then the average divergence for all of the genes should be high. The key is that we are trying to identify genes with referring articles that seem to be similar to other articles that refer to other genes.

This approach has the advantages of an article-based approach that looks at large numbers of articles. Since we are averaging the divergence contributions from each of the genes, the contribution of each gene is weighed equally. The bias towards well-studied genes is therefore mitigated. This method also focuses only on the articles that are actual references to the genes. It is a high-yield set of articles.

NDPG achieves 95% recall (18 out of 19 functional groups) at 100% precision; this is equivalent to 95% sensitivity at 100% specificity. NDPG performance is robust to different size gene groups. Smaller groups usually contain fewer genes, fewer articles, and consequently are more difficult to discover. Despite that, the NDPG is able to assign relatively high scores to these groups as well.

Figure 6.4 plots the distribution of NDPG scores for the 1900 random gene groups and the 19 functional gene groups. While there is slight overlap, most functional groups have scores that are about an order of magnitude higher than the highest score assigned to a random gene group. None of the random groups have a score exceeding 12. In Table 6.4 we have listed the scores of the individual groups of genes. Notice there is no specific pattern associated with the size of the groups, or the number of articles available.

The only adjustable parameter is the exact number of semantic neighbors that should be calculated for each article. The performance is robust to the number of neighbors; 95% recall, at 100% precision is achieved with 19, 49, or 199 neighbors. However 199 neighbors achieve the highest precision at 100% recall. At 100% recall, 199 neighbors achieve 90.5% precision, while 49 and 19 neighbors achieve 66% and 59% precision, respectively. So the performance of NDPG is superior to that of the other example coherence schemes that we have presented.

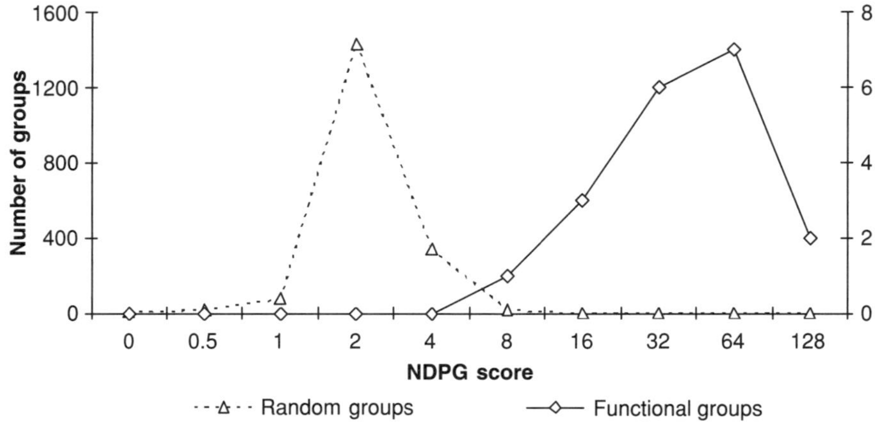

Figure 6.4 *Histogram of NDPG functional coherence scores.* Each open triangle (△) represents the count of random gene group scores in the range indicated on the horizontal axis; each open diamond (◇) represents the count of functional gene group scores in the range on the horizontal axis. The horizontal axis is in a log range. There is little overlap between the two histograms. None of the random gene groups score above 12; most of the functional gene groups score well above 12.

6.8 Corruption studies

At the beginning of this chapter we suggested that a method to assess the functional coherence of a group of genes could be used in devising groups of genes with shared data featuers that can be also explained in terms of a biological basis. Iterative optimization of features in the data alongside functional coherence scores should create groups of genes with shared biological bases. For this to be possible, the functional coherence score should (1) be able to detect imperfect groups generated by experimental data and (2) improve as groups of genes are optimized. As appropriate genes are added, and inappropriate ones are removed, the functional coherence score should improve monotonically. Only in this context does it make sense to modify groups to achieve the highest coherence score possible.

In this section we demonstrate how removing legitimate genes and replacing them with noise genes in the gene group affects the score. If the score falls off monotonically, then the score is well behaved and even partial groups have a signal. The scoring system proposed can then also be used to refine gene groups, by adding and replacing genes to increase the functional coherence score.

For two of the gene groups, *ion homeostasis* and *autophagy*, we removed genes from the gene set and swapped in random genes. We repeated this process until all but one of the original gene remained. As genes were swapped, we recalculated the NDPG score. This procedure was repeated ten times.

Table 6.4 *NDPG scores for individual functional groups.* The first column lists the functional group, the second column lists the total number of references for that group, and the final column lists the NDPG score. It is apparent that the NDPG score is not necessarily correlated with the size of the group or number of reference articles available.

GO code	Article references	NDPG
Signal transduction	3484	89.04
Cell adhesion	82	24.63
Autophagy	110	104.41
Budding	1692	112.29
Cell cycle	8399	60.61
Biogenisis	6439	40.39
Shape size control	1629	60.92
Cell fusion	2495	90.74
Ion homeostasis	667	79.73
Membrane fusion	212	18.41
Sporulation	646	10.01
Stress response	2603	32.85
Transport	4559	53.25
Amino acid metabolsim	1594	13.15
Carbohydrate metabolism	2719	29.12
Electron transport	205	19.27
Lipid metabolism	1035	94.47
Nitrogen metabolsim	264	24.06
Nucleic acid metabolism	12345	33.74

Scores slowly decrease as genes are replaced. Over half of the genes can be removed while still maintaining a strong signal (see Figure 6.5). Incomplete gene functional sets can be detected, though their scores will be less. Therefore, partial functional groups derived from experimental screens are still discernable. Figure 6.5 also suggests as NDPG scores are optimized by addition and removal of genes, more ideal functional gene groups can be obtained. There is then the possibility of using NDPG in algorithms to automatically define and redefine gene groups independently and also in the context of experimental data.

6.9 Application of functional coherence scoring to screen gene expression clusters

One goal of functional coherence assessment is to be able to rapidly evaluate experimental groups of genes created from data. One real world example of this is in gene expression array data clustering. The methods described in Section 2.4 can be used to devise many clusters of genes rapidly. The next step is to quickly examine those groups for clusters that are interesting functionally.

As a preliminary demonstration of the potential of coherence measures such as NDPG, we have scored 10 gene expression clusters that were identified manually as being biologically consistent. This is a real world test of our ability to detect meaningful groupings (Eisen, Spellman et al. 1998). Eisen and colleagues collected expression measurements on yeast genes under 79 diverse conditions. They used a hierarchical clustering algorithm to identify groups of genes with coherent gene expression patterns. Some of the gene clusters contained many genes with similar function. These clusters were manually identified and labeled with the appropriate gene function. We re-evaluated the functional coherence of these clusters automatically with NDPG. Our results are presented in Table 6.5. All clusters had very significant scores; all scores exceeded 98% of the scores of the random gene groups. So NDPG is very sensitive for these functionally coherent experimentally derived groups of genes.

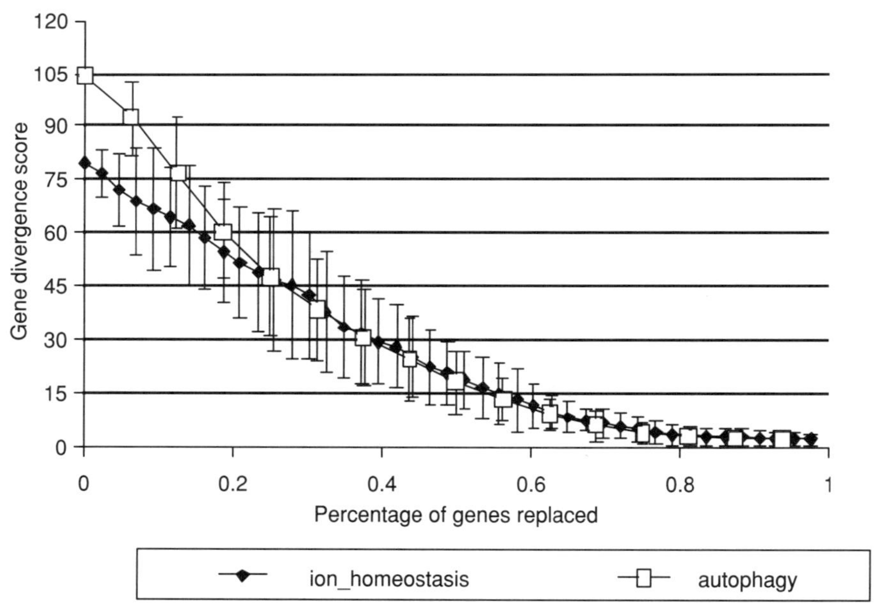

Figure 6.5 *Replacing functional genes with random genes reduces NDPG scores gracefully.* Here we have replaced genes in two functional gene groups (*autophagy* and *ion homeostasis*) with random genes, and scores were recalculated for the corrupted groups. Each point represents ten scores, error bars indicate 95% confidence interval of scores for that many genes replaced. Functional coherence scores above 12 are very significant (see Figure 6.4). Functional coherence scores remain significant despite replacement of about 56% (9 of 16 genes) of the *autophagy* genes and 58% (25 of 43 genes) of the *ion homeostasis* genes.

Table 6.5 *Assigning NDPG scores to experimentally obtained gene expression clusters.* Eisen and his colleagues clustered genes from diverse experimental conditions. They labeled ten of the clusters in Figure 2 of their paper as containing genes that represent some consistent biological function. We applied NDPG to score the functional coherence of each of these clusters. Each row represents a gene cluster. The first column lists the functional label assigned by Eisen to the gene cluster. The second column lists the number of genes in the cluster. The third column lists the functional coherence score of the cluster. The fourth column lists the percentile of the score relative to 1900 random gene groups.

Function label assigned to expression cluster (by Eisen et al.)	Number of genes	NDPG score	Score percentile
ATP synthesis	14	36.945	100.0%
Chromatin structure	8	73.072	100.0%
DNA replication	5	64.476	100.0%
Glycolysis	17	27.635	100.0%
Mitochondrial ribosome	22	5.842	98.8%
mRNA splicing	14	8.433	99.8%
Proteasome	27	154.758	100.0%
Ribosome and translation	125	55.022	100.0%
Spindle pole body assembly and function	11	10.869	99.9%
Tricarboxylic acid cycle and respiration	16	27.290	100.0%

For three of the clusters the functional coherence score was relatively poor, however, compared to the other groups. These three clusters were the only ones whose scores did not exceed the scores of all of the random gene groups. The "spindle pole body assembly and function" cluster contained 11 yeast genes; only three of these genes are among the 32 "spindle pole" genes listed by the Comprehensive Yeast Genome Database (CYGD) (Mewes, Frishman et al. 2000). Similarly, the "mitochondrial ribosome" cluster contained 22 genes; only 10 of these genes are among the 49 "mitochondrial ribosome" genes listed by CYGD. Also, the "mRNA splicing" cluster contained 14 genes; only three of these genes are among the 38 listed "mRNA splicing" yeast genes in CYGD. Many of the genes in these clusters do not represent the annotated function. While these clusters are suggestive, they are not completely coherent functional groups; they contain less than half of the genes with the reported function. Accordingly, the functional coherence scores are lower. The true functional groups of genes probably contain a coherent signal in the gene expression data, but the signal is more subtle than a clustering algorithm based on overall expression similarity can capture. Perhaps, similarity in a few particular conditions is especially critical. Detection of such subtleties will require algorithms to include functional similarity into the analysis directly; NDPG scoring may be a means to do this.

6.10 Understanding the gene group's function

NDPG determines whether a group of genes has a coherent function. It does not tell us the function. The easiest way to determine the group's function is to examine the higher scoring articles for a gene group manually or automatically. These high scoring articles are those most relevant to the group's shared function. Alternatively, keywords that describe the function of the group could be automatically determined. Investigators have already developed algorithms to find keywords in collections of biological documents that could be applied to these high scoring articles to determine function words (see Chapters 3 and 4).

References

Ashburner, M., C. A. Ball, et al. (2000). "Gene ontology: tool for the unification of biology. The Gene Ontology Consortium." *Nat. Genet.* **25**(1): 25–9.

Cherry, J. M., C. Adler, et al. (1998). "SGD: Saccharomyces Genome Database." *Nucleic Acids Res.* **26**(1): 73–9.

Eisen, M. B., P. T. Spellman, et al. (1998). "Cluster analysis and display of genome-wide expression patterns." *Proc. Natl. Acad. Sci. USA.* **95**(25): 14863–8.

Jenssen, T. K., A. Laegreid, et al. (2001). "A literature network of human genes for high-throughput analysis of gene expression." *Nat. Genet.* **28**(1): 21–8.

Mewes, H. W., D. Frishman, et al. (2000). "MIPS: a database for genomes and protein sequences." *Nucleic Acids Res.* **28**(1): 37–40.

Raychaudhuri, S., H. Schutze, et al. (2002). "Using text analysis to identify functionally coherent gene groups." *Genome Res.* **12**(10): 1582–90.

Raychaudhuri, S., H. S. Schütze, et al. (2003). "Inclusion of textual documentation in the analysis of multidimensional data sets: application to gene expression data." *Machine Learning* **52**: 119–145.

7 Analyzing large gene expression data sets

The most interesting and challenging gene expression data sets to analyze are large multidimensional data sets that contain expression values for many genes across multiple conditions. In these data sets the use of scientific text can be particularly useful, since there are a myriad of genes examined under vastly different conditions, each of which may induce or repress expression of the same gene for different reasons. There is an enormous complexity to the data that we are examining—each gene is associated with dozens if not hundreds of expression values as well as multiple documents built up from vocabularies consisting of thousands of words.

In Section 2.4 we reviewed common gene expression strategies, most of which revolve around defining groups of genes based on common profiles. A limitation of many gene expression analytic approaches is that they do not incorporate comprehensive background knowledge about the genes into the analysis. We present computational methods that leverage the peer-reviewed literature in the automatic analysis of gene expression data sets. Including the literature in gene expression data analysis offers an opportunity to incorporate background functional information about the genes when defining expression clusters. In Chapter 5 we saw how literature-based approaches could help in the analysis of single condition experiments. Here we will apply the strategies introduced in Chapter 6 to assess the coherence of groups of genes to enhance gene expression analysis approaches. The methods proposed here could, in fact, be applied to any multivariate genomics data type.

The key concepts discussed in this chapter are listed in the frame box. We begin with a discussion of gene groups and their role in expression analysis; we briefly discuss strategies to assign keywords to groups and strategies to assess their functional coherence. We apply functional coherence measures to gene expression analysis; for examples we focus on a yeast expression data set. We first demonstrate how functional coherence can be used to focus in on the key biologically relevant gene groups derived by clustering methods such as self-organizing maps and *k*-means clustering. We also demonstrate how meaningful hierarchical cluster boundaries can be defined using literature-based functional coherence measures. We then focus on extending functional coherence assessment methods to other organisms

1) Screening clusters for biological relevance
   a) Self-organizing maps
   b) k-means clustering
2) Optimizing cluster coherence to draw cluster boundaries
3) Functional coherence in different organisms
   a) Yeast
   b) Mouse
   c) Fly
   d) Worm
4) Augmenting the literature index with sequence similarity when literature is limited
5) Application of literature-based methods to a real world data set where genes have limited literature.

beyond yeast, where there is a relative paucity of literature. We comprehensively evaluate our methods to score functional coherence in four different organisms; we also discuss extending literature references with sequence similarity. Finally we demonstrate these methods on a real world data set where there is a paucity of literature.

7.1 Groups of genes

Most gene expression analysis approaches focus on defining groups of genes containing common expression features. Both clustering and classification algorithms organize expression data sets by dividing them into smaller sets of genes.

Currently, clustering methods are the most popular computational approach to apply to gene expression data. Clustering methods organize complex expression data sets into tractable subgroups, or clusters, of genes sharing similar expression patterns and thus suggesting co-regulation and possibly common biological function (Eisen, Spellman et al. 1998; Sherlock 2000). Clustering methods are reviewed in detail in Chapter 2, and countless novel clustering approaches have been purported to be excellent methods for expression data. Careful examination of the genes that cluster together can lead to hypotheses about gene function and co-regulation. But the quality of clusters and their ability to explain biological function can vary greatly.

However, clustering methods are limited in that no background knowledge is included. The prejudices of established biology may be of enormous value in the context of clustering. Published scientific text contains a distilled version of all of the most significant biological discoveries and is a potent source of functional information for analytical algorithms. This information is invaluable in guiding the investigator in understanding which clusters represent true biological phenomena. In this section we will explore the enhancement of clustering approaches with the scientific literature. At the very least, the scientific literature can be used to annotate the clusters that are obtained. A more sophisticated problem is discerning whether the cluster contains relevant biology or is spurious. In addition the scientific literature can actually be used to bias the clustering as well.

7.2 Assigning keywords

One of the first challenges of including the scientific literature in algorithms that analyze and interpret gene expression data is the assignment of key words after a group of genes with common expression features have been devised. This subject is explored in detail in Chapter 4 in the context of annotation of families of protein sequences. The same methods could be transferred to gene expression data, the only difference being that instead of families of genes with similar sequences, we have families of genes with similar expression features (Andrade and Valencia 1997). Alternatively, articles could be scored for relevance to the gene group as described in Chapter 6, and then keywords could be gleaned specifically from high scoring articles only using the same strategies.

7.3 Screening gene expression clusters

Gene expression clustering algorithms such as *k*-means and self-organizing maps and others can significantly reduce the complexity of the data that are being analyzed. For example, expression profiles for thousands of genes can be reduced to hundreds of clusters. Many of these clusters correspond to true biological phenomena, while others are the result of spurious statistical anomalies. One of the challenges of gene expression analysis is sorting through these large numbers of clusters and separating the clusters that consist of spurious genes from the functionally significant ones.

For this purpose, functional coherence scores are ideal; they are described in detail in the previous chapter. Given a group of genes, a reference index, and the text of abstracts, these methods assign a score proportional to the functional coherence of the group. Our method of choice, NDPG, works by looking at gene references, and the N most similar abstracts to them based on similarity in word use; here we used $N = 19$. NDPG scores each of the referring abstracts for overall relevance to the given gene group by counting the number of its semantically similar neighbors that also have references to group genes. The distribution of article scores is used to assess functional coherence.

NDPG can be used to do a literature-based assessment of the coherence of functional groups of genes. As we saw in the last chapter, the high scoring groups are very likely to share a biological basis. The higher the score, the more the biological significance of the genes is explained by the scientific literature. A clustering algorithm can be used to produce large numbers of groups of genes, and then NDPG can be used to screen the groups of genes that are biologically coherent. This is the manual equivalent of taking each cluster, and examining the literature to see if there is a common recurring pattern among the genes. Investigators can then focus their efforts on high scoring clusters that are most likely to be biologically relevant.

We demonstrate this concept on Eisen's yeast data set (Eisen, Spellman et al. 1998). In the last chapter we showed that the clusters that Eisen and his colleagues constructed actually have high NDPG scores. Here we used self-organizing maps to cluster the expression data collected on 2467 genes. The genes were clustered on the basis of 79 expression conditions. The genes were organized into a 15 by 15 grid of 225 clusters. Presumably clusters that are close on the grid have similar expression profiles. The number of genes in each cluster is depicted in Table 7.1(a).

Each of the clusters can be scored with NDPG. The reference index to PubMed abstracts was obtained from SGD (Cherry, Adler et al. 1998); it had references available for 2394 of the 2467 genes (97%) in the data set. There were a total of 40,351 references to 17,858 articles. Each gene had a median of eight article references, but a mean of 16.9 references. This is because the distribution of article references per gene is skewed; a few articles have many references. This is a common property of gene reference indices, and is described in greater detail in Chapter 1. This data set had the advantage of containing only genes that had excellent coverage in the scientific literature.

The NDPG scores of each of the clusters have been calculated and reported in Table 7.1(b). We regard the scores of the clusters with five or fewer genes in this case to be unreliable. Since the score is calculated on very few genes and possibly very few articles, we will disregard the scores of these small groups.

In Plate 7.1 we have displayed all of the clusters with more than five genes and with NDPG scores greater than 3. There are a total of 140 groups with more than five genes; of these only 22 have NDPG scores more than 3. Six have scores greater than 6, and four have NDPG scores more than 9. (Since a different reference index and gene set are being used here, the cutoff for a significant NDPG score is not the same as in Chapter 6.)

Table 7.1 (a) *Number of Genes in Self-Organizing Map Clusters.* Genes from the yeast data set were clustered into a 15 × 15 grid using the self-organizing map clustering method. The number of genes in each cluster is listed below in grid format. (b) *NDPG score for each cluster.* Each of the clusters in the grid was scored with NDPG. The final score cluster score is listed in grid format.

(a)

0	1	2	3	4	5	6	7	8	9	10	11	12	13	14	15
1	2	15	21	17	17	8	14	11	3	4	1	14	8	6	11
2	4	3	11	14	1	10	8	2	4	45	6	9	3	8	43
3	11	31	15	1	7	1	4	3	6	18	4	3	12	4	11
4	5	6	13	3	4	2	4	4	1	0	4	8	7	112	4
5	18	8	7	3	1	4	0	6	6	5	3	6	5	10	17
6	7	8	29	11	7	5	8	3	22	10	3	4	6	8	19
7	1	29	4	1	24	6	0	10	3	1	3	7	5	6	5
8	9	27	26	2	14	4	6	5	29	4	9	3	8	2	8
9	48	6	2	7	22	45	17	2	23	6	6	7	28	1	18
10	0	2	6	6	4	17	3	2	4	2	1	3	4	4	12
11	17	4	5	10	5	3	14	13	6	12	14	16	8	1	7
12	13	10	10	13	6	6	33	11	4	1	7	36	5	13	9
13	3	38	9	7	9	54	5	10	4	9	10	3	8	13	20
14	6	5	9	8	3	13	7	4	10	6	34	52	19	6	11
15	47	113	4	33	8	13	11	12	1	21	6	0	18	11	13

(b)

0	1	2	3	4	5	6	7	8	9	10	11	12	13	14	15
1	22.4	5.2	2.1	8.5	1.3	2.5	3.4	2.0	0.8	5.3	0.0	1.5	1.3	1.5	1.2
2	1.6	4.3	1.8	1.8	3.6	2.2	3.5	0.0	3.2	4.7	2.3	3.9	0.6	2.1	3.9
3	1.9	2.3	1.9	0.0	1.4	2.8	2.5	3.3	1.1	1.5	3.2	3.5	1.5	0.3	0.5
4	2.8	1.8	2.0	0.8	1.5	0.8	0.9	2.0	0.0	0.0	2.7	1.5	1.4	4.2	1.0
5	2.1	2.2	1.2	3.4	0.0	0.0	0.0	0.7	1.6	3.2	3.6	1.6	1.2	1.6	1.5
6	1.0	2.6	1.7	0.7	0.9	2.1	2.9	2.5	1.5	9.6	5.4	1.4	1.6	1.0	2.1
7	0.0	1.5	4.1	0.0	1.4	2.4	0.0	1.5	1.7	8.8	2.6	1.5	2.7	2.6	4.0
8	1.0	0.8	2.8	0.0	1.7	1.4	1.7	1.2	1.5	2.9	2.0	1.2	4.1	4.4	1.8
9	2.6	0.2	1.8	3.3	0.9	1.1	1.7	4.0	1.1	1.2	0.8	1.2	2.5	6.5	13.6
10	0.0	0.0	1.3	3.1	2.5	1.2	3.6	0.0	1.9	4.2	13.4	2.7	1.2	6.9	3.0
11	1.4	1.1	1.5	1.4	6.1	1.5	1.8	2.6	1.3	1.4	1.5	2.1	1.5	1.8	1.3
12	1.9	2.1	1.7	0.9	1.7	3.4	1.3	1.3	1.1	0.0	2.2	23.5	4.3	1.8	0.8
13	7.5	2.9	2.3	2.2	12.9	5.0	1.9	1.0	4.7	0.7	0.7	1.9	1.7	2.5	1.4
14	2.0	1.5	2.1	1.3	3.5	3.6	1.4	1.8	2.5	1.7	2.5	1.0	1.1	2.2	1.0
15	3.6	8.8	1.8	4.9	2.4	1.4	1.4	1.2	0.0	2.8	3.3	0.0	2.9	1.3	2.1

To independently evaluate the clusters we assess how well they correlate with Gene Ontology functional groups (see Chapter 1). For yeast, Gene Ontology (GO) annotations are well defined and can serve as a gold standard for assessment. We expect biologically relevant clusters to overlap with GO functional groups more often than not. For each GO code, we defined functional groups to contain (1) all genes with that code as an annotation and (2) all genes with a descendant of that code as an annotation. To assess the concordance or *overlap* of a cluster with a functional group we used the following formula:

$$\frac{\#(G \cap C)}{\#(G \cup C)}$$

where G is the GO functional group and C is the cluster of genes. This is the percentage of genes in either the cluster or the GO functional group that are in both. An overlap of 100% implies that the cluster and the GO functional group contain the exact same genes, whereas an overlap of 0% implies that the two groups have no genes in common. An overlap of 50% can be very significant; for example a cluster containing ten genes that shares seven genes with a functional group containing eleven genes corresponds to an overlap of 50%. An overlap of 10% can be ambiguous. For a small cluster, a 10% overlap may imply that only a single gene is in the functional group. On the other hand, given a large cluster and a very small functional group a 10% overlap may imply that the entire functional group is contained in the cluster, and it can be significant. In this chapter we evaluate clusters by finding the functional group it overlaps the best with. The expectation is that groups with high NDPG scores should have relatively high overlap with a predefined functional group.

For each of the groups we calculate the percentage overlap with all of the possible functional groups, and then select the highest overlapping group.

In Table 7.2 we list all of the clusters with NDPG scores greater than 3 and the GO functional group they overlap most strongly with. Some of the clusters have very obvious functional correlations. For example cluster B has a high NDPG score of 8.5, and has a high overlap with the *nucleosome* GO functional group of 47%. Another example is cluster L which has a high NDPG score of 13.6 which has a high overlap of 46% with the *heat shock protein* GO functional group.

It should be emphasized that while a high overlap definitely implies functional significance, a low percentage overlap does not necessarily indicate lack of functional coherence. For example, some of the clusters have a lower overlap with a GO functional group but a high NDPG score. Looking at these clusters closely may identify a common thread among these genes.

In general there is a direct relationship between NDPG and the highest percentage overlap with the GO group. Both of these measures are indirect measures of coherence of a functional group. In Plate 7.2 we have plotted the percentage overlap of each cluster as a function of the NDPG score, and

Table 7.2 *High scoring clusters.* We show all of the self-organizing map clusters with more than five genes and NDPG scores greater than 3. Groups are identified by the letter and coordinate. We list the number of genes in the cluster and its NDPG score. We also list the Gene Ontology (GO) group with which the cluster has the greatest overlap, and the percentage of that overlap.

Cluster	Coordinates	N	NDPG	GO	overlap (%)
A	1,2	15	5.2	Hydrogen transporting ATPase V1 Domain	10
B	1,4	17	8.5	Nucleosome	47
C	1,7	14	3.4	Pyrivate dehydrogenase	13
D	2,7	8	3.5	Signal peptide processing	10
E	2,10	45	4.7	Hydrogen transport	14
F	2,12	9	3.9	Copper ion transport	9
G	2,15	43	3.9	DNA strand elongation	15
H	4,14	112	4.2	Microtubule based process	15
I	6,10	10	9.6	Nitrogen starvation response	40
J	8,13	8	4.1	Hydrogen transporting ATP Synthase, central stalk	22
K	9,4	7	3.3	Isocitrate dehydrogenase	11
L	9,15	18	13.6	Heat shock protein	46
M	10,4	6	3.1	Microtubule stabilization	13
N	12,6	6	3.4	Proline metabolism	13
O	12,12	36	23.5	26 S proteosome	50
P	13,5	9	12.9	oxidoreductase, acting on the CH-NH2 group of donors, NAD/P as acceptors	20
Q	13,6	54	5	Mitochondrial ribosome	31
R	14,6	13	3.6	O Linked glycosolation	10
S	15,1	47	3.6	Nucleus	14
T	15,2	113	8.8	Cytosolic ribosome	69
U	15,4	33	4.9	Glycolysis	24
V	15,11	6	3.3	Histone acetyltransferase	13

the relationship is evident; there is a correlation between these two variables of 0.78. Consider the 22 groups with NDPG scores greater than 3. The mean overlap of these clusters with the best GO functional groups is 23%; the maximum overlap is 69% among these groups. On the other hand, let us consider the 118 genes with NDPG scores less than 3. The mean percentage overlap of these groups is 9%, and the maximum overlap is 15%. These results argue that the NDPG score is a good way to separate likely functionally coherent groups from the rest.

This same strategy could be applied just as easily to any other clustering method, such as *k*-means clustering, that produces a set of disjoint clusters. In practice NDPG provides a means of incorporating information from the scientific text in the analysis of gene expression data. After a clustering method is applied, we can use text analysis in the form of NDPG to prioritize the clusters and select the biologically meaningful ones. These clusters allow us to characterize the key biological elements in a data set. They also open the possibility of assigning a function to uncharacterized genes. An uncharacterized gene in a highly coherent cluster likely shares that same function.

Furthermore, NDPG scoring offers an objective means to assessing how appropriate a clustering method is for a data set. The performance of a clustering method can be measured for a given data set, by examining the NDPG scores of the clusters produced by it. A high quality method will generate more high scoring clusters.

7.4 Optimizing cluster boundaries: hierarchical clustering

Not only can literature-based coherence scores such as NDPG be used to identify the key clusters, they can also be used to actually define cluster boundaries. Defining cluster boundaries such that the NDPG scores are maximized insures high quality clusters that are less subject to the whims of statistical noise. We propose drawing boundaries so that the group of genes is coherent with that which is published in the scientific literature.

The most commonly used clustering method, hierarchical clustering, offers considerable ambiguity in determining the exact cluster boundaries. Hierarchical clustering organizes expression data into a binary tree, or dendrogram, in which the leaves are genes, and the interior nodes (or branch points) are candidate clusters. The more similar the gene expression patterns of two genes, the closer they are within the tree structure. In many cases, genes with a shared biological function also share expression features and therefore cluster together in a node. Hierarchical clustering is described in greater detail in Section 2.4.

Once a dendrogram has been devised, the challenge is to define the final cluster boundaries by pruning the tree (see Plate 7.3). In other words, the challenge is to select nodes appropriately so that the genes are divided into non-overlapping biologically meaningful clusters. Typically, cluster boundaries are drawn so that the final clusters contain functionally related genes. In practice, investigators define clusters by tedious application of their own knowledge of these genes. Some have proposed automatically selecting nodes and defining boundaries based on statistical properties of the gene expression profiles within them; however the same statistical criteria may not be generally applicable to identify all relevant biological functions. The result is that the boundaries do not necessarily divide the dendrogram by biological function at all, and can appear arbitrary to the eye of a biologist.

After application of hierarchical clustering to the given gene expression data set, text about genes can instead be used to resolve hierarchical cluster boundaries to correspond to biological functions.

Application of hierarchical clustering on K genes yields $K - 1$ internal nodes containing at least two genes and K leaf nodes containing a single gene. The root node contains all K genes. The goal of the algorithm presented here is to "prune" the dendrogram, or rather to select a subset of nodes, S, such that each gene is contained in a single selected node (Plate 7.3). The objective of our pruning strategy is to maximize the functional relatedness of the genes in each selected node based on the scientific literature. We propose selecting nodes so that the total weighted average of NDPG cluster scores is maximized (Raychaudhuri, Chang et al. 2003). Nodes are weighted by their size, so larger, but equally coherent groups have an advantage. The selected nodes with the highest scores are likely to constitute functionally relevant clusters. As in the previous section, the NDPG scores can be used to prioritize clusters for subsequent detailed manual analysis and experimental follow-up.

The NDPG weighted average of a disjoint set of nodes S is defined as:

$$F_S = \frac{1}{K} \sum_{i \in S} n_i \cdot f_i \tag{7.1}$$

where f_i is the NDPG score of the node i and K is the total number of genes. The average is weighted by the number, n_i, of genes in the node i. The goal of our algorithm is to select disjoint nodes S and prune the tree so that F_s is maximized. The key insight to the algorithm is that if a node is in the optimal set, then the NDPG score of the node must exceed the weighted average NDPG score of any disjoint subset of its descendants. That is for any node i contained in the optimal set S,

$$n_i \cdot f_i \geq \sum_{j \in S_i} n_j \cdot f_j \tag{7.2}$$

where the nodes S_i are any disjoint set of descendent nodes of node i. Were this not the case for a given subset of a node i, S_i, then node i could be replaced by the nodes in S_i in the final set of optimal nodes, and the weighted NDPG score calculated with equation (7.1) would be increased.

We use this very property to identify the key nodes. We start at the bottom of the tree and assume that all of the leaf nodes constitute the set that gives the largest average score. We then work our way up the tree, replacing those nodes with new nodes that satisfy equation (7.2).

The algorithm is summarized in Table 7.3. Our algorithm has three states that a node can be in: *unvisited*, *visited*, and *selected*. After running the algorithm, the set of *selected* nodes constitute the final set S of clusters; the remainder of the nodes will be in the *visited* state. Initially all internal nodes are *unvisited* and the terminal leaves are *selected*. The pruning algorithm proceeds iteratively, visiting *unvisited* nodes whose descendents are in the *visited* or *selected* state; the status of the node is changed to *visited*. If the functional coherence score of this node equals or exceeds that of the weighted average of its *selected* descendants, it is placed in the *selected* state, and all of its *selected* descendant's children are de-*selected* and placed in the *visited* state. The process repeats until all nodes up to the root node have been examined; the nodes that are still *selected* define the final set of clusters that maximize the NDPG weighted average across the hierarchical tree.

Since our goal is to identify small specific coherent groups, we assign groups containing more than 200 referenced genes a score of zero. Also NDPG might be unreliable for extremely small groups of genes, so groups with fewer than six referenced genes are assigned a score of zero.

To test this approach, we apply our pruning method to the *Saccharomyces cerevisiae* (yeast) gene expression data set used in the last section (Eisen, Spellman et al. 1998). If our method is successful, the expression clusters defined by our method should correspond to well-defined functional groups of genes. As in the previous section, Gene Ontology (GO) is available for use as a gold standard for comparison. We again use the reference index from the *Saccharomyces Genome Database* (SGD) (Cherry,

Table 7.3 *Algorithm to prune gene expression dendrogram into disjoint clusters.* The NDPG score of a node i is represented as f_i, the number of nodes in the cluster is n_i. The set of descendants of a node i in the selected state is $Sel(i)$. Nodes are in one of three states, *selected, unvisited,* or *visited.* Initially all internal nodes are *unvisited* and the terminal leaves are *selected.* After running the algorithm, the nodes that are *selected* define the final set of disjoint clusters.

1.	For each node i, determine n_i *and* f_i
2.	Assign all nodes state *unvisited*
3.	Assign leaf nodes state *selected*
4.	While there exists *unvisited* nodes
5.	For each node i (1) in the *unvisited* state and (2) with both children in state *visited* or *selected*
6.	Assign node i state *visited*
7.	If $n_i \cdot f_i \geq \sum_{j \in Sel(i)} n_j \cdot f_j$
8.	Assign node i state *selected.*
9.	Assign all nodes in *Sel(i)* state *visited*
10.	Nodes in state *selected* define cluster boundaries

Adler et al. 1998). Our results are comparable to those produced manually by the original investigators and required only minutes of computation.

We use the gene expression analysis software *Cluster* to create a hierarchical clustering of the yeast gene expression data. To create the clustered dendrogram of the data we use *average-linkage clustering* with the *centered correlation* metric option to calculate distances. In inter-gene distance calculations, conditions are differentially weighted according to the scheme introduced in the original publication; each condition is weighted with the square root of the number of conditions in that series.

Hierarchical clustering of the yeast gene expression data set creates a total of 2466 internal nodes containing two or more genes; availability of the SGD literature reference index and corpus of article abstracts allows NDPG evaluation of the functional coherence of each node. We use the overlap with GO functional groups as an independent measure of functional coherence. Figure 7.1(a) shows that the literature-based NDPG score of a node predicts how well it corresponds with a GO functional group (non-parametric Spearman rank correlation $r = 0.81$). Therefore, selecting nodes with large NDPG scores will result in selecting nodes whose genes share a common function.

Defining cluster boundaries that respect biological function by maximizing total NDPG weighted average selects 369 non-overlapping nodes as the final clusters. These nodes are indicated as black circles in Figure 7.2(a). Figure 7.2(b) individually plots the node corresponding to the function *threonine endopeptidase*. The other points in this plot correspond to other nodes that are either ancestors or descendants of that node; these nodes contain a subset or superset of the genes in the node. The selected node has the greatest concordance with any GO functional group than all of the other nodes in that plot; these are nodes that might have been selected instead. This figure suggests that the tree was in fact pruned as best as it could have been, and that a node at exactly the right level was selected.

We ranked the top 20 clusters by NDPG scores in Plate 7.4. To evaluate whether the selected genes are true functional groups of genes, we checked the degree to which they corresponded to any of the functional groups defined by GO. Listed alongside the clusters is the best corresponding GO code and a graphical depiction of the overlap between that GO code and the cluster. Nine of the ten functional clusters noted in the original publication of the data set are included in our list along with other functional clusters. These functions include *threonine endopeptidase*, *ATP synthesis coupled proton transport*, *ATP dependent DNA helicase, nucleosome*, *electron transport, glyceraldehyde 3-phosphate dehydrogenase, cytosolic ribosome*, *mitochondrial ribosome*, and the *tricarboxylic acid cycle*. The other depicted groups also contain functionally related genes, but were not described in the original publication, such as *pheromone response*, *heat*

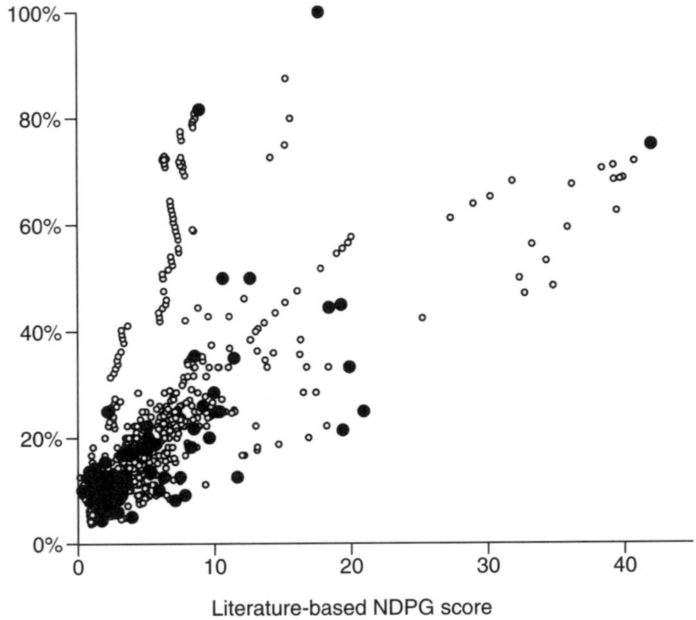
100%
80%
60%
40%
20%
0%
0
10
20
30
40
Literature-based NDPG score

(a)

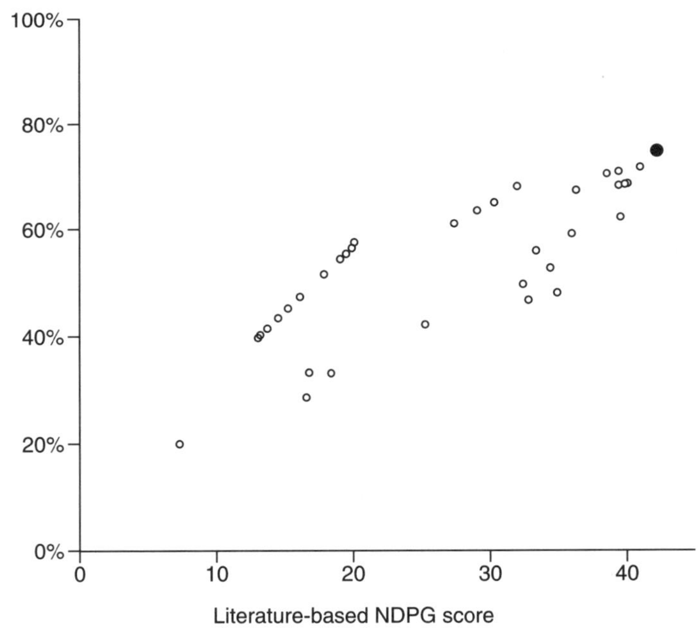
100%
80%
60%
40%
20%
0%
0
10
20
30
40
Literature-based NDPG score

(b)

shock protein, and *nucleolus*. These are other potentially interesting clusters that our method picked up easily in a first pass.

As noted previously, the percentage overlap can underestimate the functional relatedness of a gene group. For example, the eleventh listed cluster has the highest overlap with the *glyceraldehydes-3-phosphate dehydrogenase* GO code, but the non-G3PD genes in the cluster are other closely related *glycolysis* genes.

Hierarchical clustering can be implemented in multiple different ways (such as average linkage, centered linkage, etc.) with one of a wide array of metrics (such as Euclidean, Manhattan, jack-knife, etc.) and weighting schemes. In this study we did not wish to explicitly evaluate the choice of hierarchical clustering implementation. We attempted to use a methodology that was as consistent as possible with the original publication so that our results were comparable. However, maximization of the NDPG weighted average to select cluster boundaries could be used in evaluating the output of different implementations of hierarchical clustering and selection of the best one. The better implementation will produce hierarchical trees that are more easily segmented into clusters that respect biological function. Such hierarchical trees will have a higher total maximized NDPG weighted average than trees produced by an implementation less effective for the specific data set.

The most labor-intensive component of gene expression array projects is the identification of biologically relevant clusters and optimization of cluster boundaries. This task is difficult and often arbitrary, requiring laborious steps of gathering information on genes within a cluster, identifying a common biological process, and drawing a boundary line somewhere around a cluster. Practically, it amounts to identifying the interesting aspects of a large experiment. This method not only automates the identification of biologically relevant data using the same source literature that researchers would access to make the same comparisons by hand, but it also creates an optimized version of each cluster, at the level of highest enrichment for a given biological function.

Figure 7.1 *Functional coherence increases with NDPG score.* (a) After clustering the yeast gene expression data into a dendrogram with 2466 individual nodes, we plotted the literature-based *NDPG* score of the 1150 nodes containing 6–200 genes on the x-axis and the highest percentage concordance with a GO functional group on the y-axis. Black circles indicate the nodes selected by the computational method. There is a correlation of $r = 0.81$ between these two variables. (b) To demonstrate that the pruning algorithm selects the most appropriate node, we plot the *threonine endopeptidase* cluster, and its ancestors and descendents in the hierarchical clustering. We have plotted the *NDPG* score and the highest percentage concordance with a GO functional group for these nodes. The selected cluster is represented with a black circle. The ancestors and descendants nodes represent clusters that are subsets or supersets of the genes in the selected node. Notice that the selected node has optimal overlap. In other words, it is the best of the possible clusters that could have been selected.

7.5 Application to other organisms besides yeast

The examples we have provided so far have been focused on yeast data; a very high percentage of the genes in those data sets had literature references. The literature in this organism is extremely comprehensive, and most genes have some documentation. In addition the total quantity of literature, and the quality of the data sets is very high. For most other organisms, however, there is a relative paucity of literature. Most genes are not described at all. The available literature indices are poorer. In addition the scope of the literature can often be narrow. For example, a large proportion of the literature on *drosophila* is focused on developmental biology, while investigation of molecular biology and biochemistry is limited. In theory, the application of literature-based approaches to gene expression data from high-level organisms is no different from application to yeast data. In practice, it can be challenging since the literature may be limited. The first step is to assess the performance of literature-based metrics in different organisms.

To get a sense of the performance of NDPG across different organisms, we evaluate NDPG on 2796 diverse functional groups generated by the Gene Ontology consortium in four organisms (Raychaudhuri and Altman 2003). In the last chapter, a small-scale, preliminary evaluation conducted on 19 yeast functional groups, NDPG achieved 95% sensitivity at 100% specificity.

We use Gene Ontology (GO), which was devised to describe genetic function in a standard way across many types of organisms, as our gold standard. To evaluate our method we assembled functional groups of genes from four species: *Saccharomyces cerevisiae* (yeast), *Mus muscularis* (mouse), *Drosophila melanogaster* (fly), and *Caenorhabditas elegans* (worm).

We used the GO annotations of genes to define functional groups. Each GO term generates a species-specific gene group consisting of genes assigned that term. Random groups of genes were assembled also. Ideally, we expect that the functional groups will receive high scores and that random groups will receive low scores. The better NDPG is at separating the random and functional groups, the more effective we expect it will be in that organism.

For each of the four organisms, Gene Ontology assignments were obtained from the genome databases for each organism: SGD for yeast (Cherry, Adler et al. 1998), MGD for mouse (Blake, Richardson et al. 2002), FlyBase for fly (Gelbart, Crosby et al. 1997), and WormBase for worm (Stein, Sternberg et al. 2001)). The assignments are either explicitly indicated, or are inferred from assignments to more specific terms. Terms ascribed by the database are explicit annotations. Since GO is hierarchical, a specific annotation implies more general annotations that are parents in the hierarchy; we refer to these as inferred annotations. For each species, a Gene Ontology term that had six or more genes assigned that term defined a functional group. There were a total of 2796 such functional groups across all four species. Each group was scored with NDPG. For each organism 200 random groups were also devised of each size: 6, 12, 24, 48, and 96 genes, for a total of 1000 random groups per organism.

Reference indices were assembled for each species by collecting gene references from the same online resources. A gene reference list was obtained from each of the genomic databases. All relevant article abstracts were obtained from PubMed.

Table 7.4 contains descriptive statistics about the literature index for each organism and the GO annotations also. Mouse has the highest number of genes with references, while worm has the fewest. For the total number of references per article and references per gene, the mean exceeds the median in all organisms. For each organism there are a few outlier genes with many article references, and there are a few outlier articles with many gene references.

All of the different organisms used about the same number of GO codes for annotation, about 2500 codes, except worm which used fewer, about 500 codes. For each term assigned explicitly to a gene as an annotation, many more annotations that are parents of the term are also implied. The more specific the term, the more inferred terms apply to the same gene. In general the yeast and fly annotations are very specific, and the ratio of inferred annotations to explicit annotations is large (3.6 and 3.3) compared to mouse and worm, where the annotations are more general (2.5 and 2.2).

The NDPG method was used to score all random groups; the 99.9th percentile of scores for each organism was identified as a cutoff. A separate cutoff was generated for each organism; the cutoffs are listed in the table. The cutoff can be thought of as a 99.9% specificity threshold. The probability that a random group will score above the cutoff is less than 0.001.

The functional groups derived from GO were scored with NDPG. The NDPG method is effective at identifying groups of genes that are functionally coherent in multiple organisms. The percentage of functional groups within the organism scoring above the cutoff is tabulated in Table 7.5. This percentage of groups above the cutoff is the sensitivity of our method. The results are presented separately for each of the three branches of GO. For all GO branches yeast has the greatest percentage of groups exceeding the cutoff, followed by mouse, then fly, then finally worm. The method achieves 96%, 92%, 82%, and 45% sensitivity at 99.9% specificity in yeast, mouse, fly, and worm, respectively. When all worm functional groups are combined, the median group score of worm functional groups is less than the cutoff.

Table 7.4 *Summary of literature index and GO groups for NDPG evaluation across four organisms.* For each of the four organisms we have listed summary statistics for the reference lists obtained from the genome databases and GO annotations. The top half of this table summarizes the reference lists obtained from the genome centers. This table lists the total number of genes that have references and the total number of articles referenced for each of the four organisms. Also listed are the total number of references, that is the number of links between genes and documents. The lower half of this table summarizes the Gene Ontology annotations obtained from the GO consortium. Most organisms use only a few of the 4773 process, 977 function, and 5015 function GO codes. Explicit GO annotations are ones assigned directly by GO, while inferred annotations are more general annotations that are implied by explicit annotation of a more specific code. The ratio indicates the average number of inferred parents terms generated by each explicit term annotation. The greater the ratio, the deeper the average annotation is down the hierarchical tree of GO.

		Yeast	Mouse	Fly	Worm
Genes with reference		5151	26148	14732	2289
Articles		22934	41669	15495	2144
References		62473	113738	97117	13659
References/article	median	2	1	3	4
	mean	2.73	2.73	6.27	6.37
References/gene	median	4	1	1	2
	mean	12.12	4.35	6.59	5.97

		Yeast	Mouse	Fly	Worm
Genes with codes assigned		4276	6148	4042	523
GO codes	process	874	904	1019	196
	component	251	233	347	42
	function	1132	1432	1458	246
	total	2257	2569	2824	484
Explicit GO annotations		13770	27122	14405	2235
Inferred GO annotations		49781	68075	47801	5017
Ratio explicit/implicit		3.62	2.51	3.32	2.24
Annotations/gene	median	14	15	14	13
	mean	14.86	15.48	15.39	13.87
Annotations/code	median	3	3	2	3
	mean	28.15	37.06	22.02	14.98

The variable performance in the four different organisms can be accounted for by different factors. The first is the quality of the references in the reference indices; in a good reference index, genes should be connected to the appropriate articles. This is difficult to objectively assess. It is possible that different literature indices have more or less appropriate references depending on the articles available and the level of care placed in putting the literature resource together.

Table 7.5 *Sensitivity of NDPG in different organisms.* The cutoff score is defined so that the NDPG score of 99.9% of random groups in that organisms scores less than the cutoff. In each of the organisms and each of the three GO categories we have calculated the percent of functional groups that exceed the cutoff score.

		Yeast	Mouse	Fly	Worm
Process codes	Number of groups	429	354	349	71
	Median group size	21	20	16	17
	Median NDPG score	15.3	10.2	5.2	1.4
	% of groups exceeding cutoff	97.4%	87.9%	86.8%	46.5%
Component codes	Number of groups	148	111	151	18
	Median group size	20	18	16	16
	Median NDPG score	18.7	11.7	5	2.4
	% of groups exceeding cutoff	94.6%	91.0%	81.5%	77.8%
Function codes	Number of groups	264	435	382	84
	Median group size	17	16	17	15
	Median NDPG score	11.4	13.4	3.6	1.5
	% of groups exceeding cutoff	93.6%	96.1%	78.3%	36.9%
All codes	Number of groups	841	900	882	173
	Median group size	20	18	16	16
	Median NDPG score	15.1	11.8	4.5	1.6
	% of groups exceeding cutoff	95.7%	92.2%	82.2%	45.1%
99.9% specificity cutoff		3.43	3.19	1.34	1.63

A second issue is the abundance of available articles in the reference index. Yeast has the strongest performance; it has over 20,000 articles and a 4:1 ratio of articles to genes. Worm, on the other hand, has the smallest corpus with one-tenth the number of articles that the yeast reference index has and a ratio less than 1:1 of articles to genes; our method is less than half as sensitive for worm functional groups. For application to gene expression this could be a severely limiting factor.

An additional contributing factor may be the quality of the GO annotations themselves. Gene Ontology is a massive effort, and remains a work in progress. Since this is the case, it is perhaps not an ideal gold standard yet. Currently annotation of genes with GO codes remains an active area with many groups experimenting with different strategies involving manual and computational analysis of literature, sequence, and experimental data (Hill, Davis et al. 2001; Hvidsten, Komorowski et al. 2001; Dwight, Harris et al.

2002; Raychaudhuri, Chang et al. 2002; Schug, Diskin et al. 2002; Xie, Wasserman et al. 2002). The on-line resources for the different organisms rely more heavily on different strategies of annotation. The strategy used to make a specific annotation is listed as an "evidence code". (See Chapter 8 for a more detailed discussion of evidence codes.) We considered IDA ("inferred from direct assay") and TAS ("traceable author statement") as the two highest quality and most reliable evidence codes. We determined the percentage of inferred and explicit annotations that could be attributed to each of these two evidence codes in the three GO branches of the four organisms. In Figure 7.2 it is evident that there is a relationship between the percentage of high quality annotations and our method's performance. The percentage of high quality annotations is an indication of the amount of manual effort involved in that organism's GO annotation. We reason that the more effort, the better the quality of the annotation, and the more reliable a gold standard it is, and consequently the better our performance.

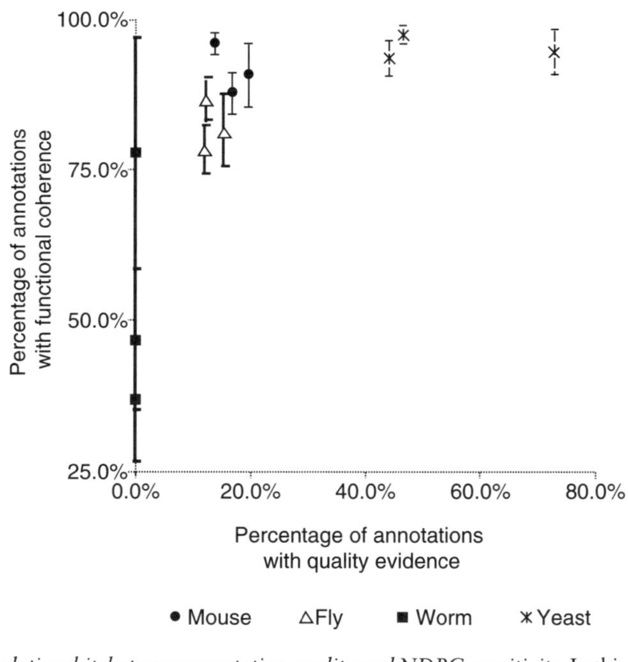

Figure 7.2 *Relationship between annotation quality and NDPG sensitivity.* In this figure we have plotted the percentage of annotations attributable to either the TAS ("traceable author statement") or IDA ("inferred from direct assay") annotations for each GO branch and organisms. These annotations are generally regarded as reliable, high quality annotations. The resulting percentages were plotted against the percentage of functional groups that *NDPG* was able to successfully identify in that GO branch and organism. In general, annotated sets where many of the annotations were derived from high quality evidence had a greater percentage of annotated groups exceeding the cutoff.

A functional group may not be identified as functionally coherent if the shared function is not represented in the corpus of scientific literature. This may be the case if the function has not yet been described in the literature, or if the function has not been well studied in that organism. For example the *tricarboxylic acid cycle* (TCA) functional group in yeast receives a significant score of 15.43, whereas in mouse the same functional group receives an insignificant score of 1.97. The subject of TCA genetics has not been well described in the mouse literature. A MedLine query for articles assigned the MeSH subject headings of “tricarboxylic acid cycle” and “genetics” yielded 365 articles. Only 13 of these articles had the “mouse” MeSH heading also, and none of those 13 references were listed in the mouse reference index. In contrast, 52 had the “yeast” MeSH heading and of those 32 were listed in the yeast reference index. The TCA GO annotations in mouse were made without direct reference to the literature. Eight of the nine genes were assigned the TCA function because of the presence of an appropriate keyword in the SWISS-PROT sequence entry for the gene; these annotations were assigned the evidence code IEA (“inferred from electronic annotation”). The other gene was assigned the TCA function by sequence similarity search; this annotation was assigned the evidence code “ISS” (“inferred from sequence similarity”). Since NDPG is an approach based on primary literature, this functional group is missed altogether. This issue might be mitigated if references to homologous genes in other organisms were included as additional references to genes. Undiscovered functions would still not be discernible.

These data suggest that there is quite a bit of power in the literature for these different organisms, and that while NDPG is somewhat less effective outside of yeast, it is still quite powerful and can detect most functional groups.

7.6 Identifying and optimizing clusters in a *Drosophila* development data set

We apply our hierarchical pruning method to a different gene expression data set with very different properties. In this more challenging test, we applied this strategy to analyzing a *Drosophila melanogaster* (fly) development series containing expression measurements for 4040 expressed sequence tags. The fly gene expression data set consisted of measurements from over 84 conditions, 75 of which were part of a wild type developmental time series, four that were segregated by sex, and five that involved mutations in specific genes. These expressed sequence tags corresponded to sequences for 3987 unique genes (Arbeitman, Furlong et al. 2002). However, the available reference index from FlyBase contained PubMed references for only 1681 of the 3987 unique fly data set genes represented in the data set. This data set is more challenging since a minority of the genes have been studied and have any primary literature.

When we evaluated NDPG it was 96% sensitive in yeast and 82% sensitive in fly at discriminating between functional groups of genes and random groups of genes at 99.9% specificity. We also found that one of the limitations of this (and likely any) literature-based approach is that certain biological functions have not been studied and reported on in the literature in certain organisms. For example, cellular and metabolic functions of many genes are better characterized in yeast than in fly or mouse.

So, in this case transferring references from well-studied homologous genes from other model organisms as described in detail in Chapter 4 is crucial. There simply are few genes that have the necessary references.

As an example consider our discussion in the last section about the function *tricarboxylic acid cycle* (TCA), which was a well-studied function in yeast. So an unstudied TCA fly gene could be assigned appropriate and informative references if we transferred the references from the corresponding well-studied homologous yeast genes.

We use sequence similarity searches to identify homologous genes for each gene in the study, and associate references from the homologous gene to the study gene. We were able to associate 3962 fly data set genes with protein sequences from SWISS-PROT or SPTREMBL. We then constructed a database of all of the genes in fly, yeast, and mouse with five or more PubMed references assigned by Flybase, SGD, or the Mouse Genome Database (MGD). We obtained protein sequences for all of these genes from the same databases. Then, for each of these 3962 fly data set gene protein sequences, BLAST was used to find the single most similar well-documented protein sequence corresponding to a fly, yeast, or mouse gene. The fly gene was assigned references from the most similar gene with more than four references only if the e-value score of the similarity was less than 1×10^{-6}. This is a very liberal cutoff. We did not transfer references if the e-value was larger than this arbitrary threshold, as the similarity may have represented a local or spurious similarity.

The initial literature reference index obtained from FlyBase contained primary references for 1681 of the 3987 genes (42%) in the data set. There were a total of 30,622 references to 11,070 articles. Each gene had a median of three article references and a mean of 18.2 references. In the augmented reference index, containing references transferred from homologous genes, 2602 of the 3987 genes (65%) had references. There were a total of 77,509 references to 29,115 articles. Each gene had a median of eight article references and a mean of 29.8 references.

We apply the strategy presented in Section 7.4. To create the clustered dendrogram of the data we use *average linkage clustering* with the *uncentered correlation* metric option to calculate gene distances. Defining cluster boundaries by maximizing NDPG weighted average selects 525 non-overlapping nodes as the final clusters. Many of the defined clusters correspond to well-defined biological functions such as photoreceptor genes, protein degradation, protein synthesis, muscle function, citric acid cycle, and proton transport (Table 7.6). Some of these clusters listed are graphically depicted in Plate 7.5. Most of these clusters correspond exactly or closely to clusters described in the original publication of the data. These are discussed in detail, and validated with in situ hybridization and mutation experiments in that publication.

One novel cluster represents uncharacterized maternally expressed genes that localize primarily to the nucleolus; this functional cluster was not identified in the original publication and has the highest NDPG score of the selected nodes (Plate 7.5a). The maternal expression of these genes is

Table 7.6 *Fly functional clusters*. Functional clusters obtained after using NDPG to define boundaries on a hierarchical clustering of a fly development time series. Here we list the top 20 clusters sorted by NDPG score; the score and the number of genes in the cluster are in the first and second column. We also list an appropriate function if it was immediately apparent.

NDPG score	N	Function
22.5	7	Nucleolar maternally expressed
20.5	7	Vacuolar ATPase
8.3	7	Photoreceptor
6.7	41	Proteasome
6.6	8	Vacuolar ATPase
6.5	7	T-ring complex
6.0	10	TCA cycle
5.2	7	Cell adhesion
5.0	34	Ribosomal
4.8	7	Vesicle transport – coatomer
4.8	12	
4.1	9	Muscle
4.1	13	
3.9	7	
3.7	22	Strict maternal
3.7	7	Photoreceptor
2.9	10	
2.7	12	
2.7	12	
2.7	12	

apparent from the expression profile: transcripts are seen in the female adult, but not the male adult, and in the embryo. These genes likely constitute an interesting biological module of developmental regulation in fly genes. Only two genes in the cluster are well studied, each with five primary papers listed in FlyBase. It has already been demonstrated that these two genes, the Fbgn0029196 (Nop5) and FBgn0023184 (Nop60B) genes, are in fact maternally expressed genes that localize to the nucleolus (Phillips, Billin et al. 1998; Vorbruggen, Onel et al. 2000).

Without the augmented reference index, clusters such as the muscle cluster are identified, as muscle development is extensively studied in fly. Many of the clusters that contain genes where the functional commonality is biochemical or molecular in nature, such as ribosomal genes, are missed without the augmented reference index. Most biochemical aspects of genes have not been explored in detail in fly; rather homologs in yeast and other species have been carefully studied. In fact, proper resolution for approximately half of the labeled functional clusters, including the nucleolar maternal cluster in Plate 7.4(a), required the use of the augmented reference index, as the published primary literature on the fly genes was sparse.

One of the primary goals of gene expression analysis is to attribute functions to unidentified genes and identify novel functions based on gene co-expression. If a gene with unknown function is in a functionally coherent cluster, it likely shares the common function of the other genes in the cluster. Experimental follow-up is necessary to confirm the putative gene function. In addition, detailed examination of unstudied genes just outside the cluster may be fruitful since they may also share the cluster function.

For example, Plate 7.4d appears to be a cluster of muscle genes. Some of the genes have not been specifically annotated as muscle expressed genes, but are likely candidates. Glycogenin, Fbgn0034603, was recently confirmed to be a muscle specific gene with in situ hybridization (Arbeitman, Furlong et al. 2002).

Not only has this method almost completely recapitulated the biologically relevant associations found through months of hands-on, one-gene-at-a-time work by teams of scientists working in both yeast and *drosophila*, but it has also been able to identify new clusters that were missed by the primary researchers (personal communication, Farhad Imam, 2003). Furthermore, this method was able to accomplish this task on the order of hours! Text-based approaches therefore other researchers the potential to reduce significantly the amount of data analysis required to begin to make meaning from the mountain of experimental data.

References

Andrade, M. A. and A. Valencia (1997). "Automatic annotation for biological sequences by extraction of keywords from MEDLINE abstracts. Development of a prototype system." *Proc. Int. Conf. Intell. Syst. Mol. Biol.* **5**(1): 25–32.

Arbeitman, M. N., E. E. Furlong, et al. (2002). "Gene expression during the life cycle of Drosophila melanogaster." *Science* **297**(5590): 2270–5.

Blake, J. A., J. E. Richardson, et al. (2002). "The Mouse Genome Database (MGD): the model organism database for the laboratory mouse." *Nucleic Acids Res.* **30**(1): 113–5.

Cherry, J. M., C. Adler, et al. (1998). "SGD: Saccharomyces Genome Database." *Nucleic Acids Res.* **26**(1): 73–9.

Dwight, S. S., M. A. Harris, et al. (2002). "Saccharomyces Genome Database (SGD) provides secondary gene annotation using the Gene Ontology (GO)." *Nucleic Acids Res.* **30**(1): 69–72.

Eisen, M. B., P. T. Spellman, et al. (1998). "Cluster analysis and display of genome-wide expression patterns." *Proc. Natl. Acad. Sci. U S A.* **95**(25): 14863–8.

Eisen, M. B., P. T. Spellman, et al. (1998). "Cluster analysis and display of genome-wide expression patterns." *Proc. Natl. Acad. Sci. U S A.* **95**(25): 14863–8.

Gelbart, W. M., M. Crosby, et al. (1997). "FlyBase: a *Drosophila* database. The FlyBase consortium." *Nucleic Acids Res.* **25**(1): 63–6.

Hill, D. P., A. P. Davis, et al. (2001). "Program description: Strategies for biological annotation of mammalian systems: implementing gene ontologies in mouse genome informatics." *Genomics* **74**(1): 121–8.

Hvidsten, T. R., J. Komorowski, et al. (2001). "Predicting gene function from gene expressions and ontologies." *Pac. Symp. Biocomput*: 299–310.

Phillips, B., A. N. Billin, et al. (1998). "The Nop60B gene of *Drosophila* encodes an essential nucleolar protein that functions in yeast." *Mol. Gen. Genet.* **260**(1): 20–9.

Raychaudhuri, S. and R. B. Altman (2003). "A literature-based method for assessing the functional coherence of a gene group." *Bioinformatics* **19**(3): 396–401.

Raychaudhuri, S., J. T. Chang, et al. (2003). "The computational analysis of scientific literature to define and recognize gene expression clusters." *Nucleic Acids Res.* **31**(15): 4553–60.

Raychaudhuri, S., J. T. Chang, et al. (2002). "Associating genes with gene ontology codes using a maximum entropy analysis of biomedical literature." *Genome Res.* **12**: 203–214.

Schug, J., S. Diskin, et al. (2002). "Predicting gene ontology functions from ProDom and CDD protein domains." *Genome Res.* **12**(4): 648–55.

Shatkay, H., S. Edwards, et al. (2000). "Genes, themes and microarrays: using information retrieval for large-scale gene analysis." *Proc. Int. Conf. Intell. Syst. Mol. Biol.* **8**(10): 317–28.

Sherlock, G. (2000). "Analysis of large-scale gene expression data." *Curr. Opin. Immunol.* **12**(2): 201–5.

Stein, L., P. Sternberg, et al. (2001). "WormBase: network access to the genome and biology of Caenorhabditis elegans." *Nucleic Acids Res.* **29**(1): 82–6.

Vorbruggen, G., S. Onel, et al. (2000). "Restricted expression and subnuclear localization of the Drosophila gene Dnop5, a member of the Nop/Sik family of the conserved rRNA processing factors." *Mech. Dev.* **90**(2): 305–8.

Xie, H., A. Wasserman, et al. (2002). "Large-scale protein annotation through gene ontology." *Genome Res.* **12**(5): 785–94.

8 Using text classification for gene function annotation

Recognizing specific biological concepts described in text is an important task that is receiving increasing attention in bioinformatics. To leverage the literature effectively, sophisticated data analysis algorithms must be able to identify key biological concepts and functions in text. However, biomedical text is complex and diverse in subject matter and lexicon. Very specialized vocabularies have been developed to describe biological complexity. In addition, using computational approaches to understand text in general has been a historically challenging subject (Rosenfeld 2000). In this chapter we will focus on the basics of understanding the content of biological text. We will describe common text classification algorithms. We demonstrate how these algorithms can be applied to the specific biological problem of gene annotation. But text classification is also potentially instrumental to many other areas of bioinformatics; we will see other applications in Chapter 10.

There is great interest in assigning functional annotations to genes from the scientific literature. In one recent symposium 33 groups proposed and implemented classification algorithms to identify articles that were specifically relevant for gene function annotation (Hersh, Bhuporaju et al. 2004). In another recent symposium, seven groups competed to assign Gene Ontology function codes to genes from primary text (Valencia, Blaschke et al. 2004). In this chapter we assign biological function codes to genes automatically to investigate the extent to which computational approaches can be applied to identify relevant biological concepts in text about genes directly. Each code represents a specific biological function such as "signal transduction" or "cell cycle".

The key concepts in this chapter are presented in the frame box. We introduce three text classification methods that can be used to associate functional codes to a set of literature abstracts. We describe and test maximum entropy modeling, naive Bayes classification, and nearest neighbor classification. Maximum entropy modeling outperforms the other

1) Gene function vocabularies
   a) Gene Ontology
   b) Enzyme Commission
   c) Kyoto Encyclopedia of Genes and Genomes
2) Text classification
   a) Maximum entropy
   b) Nearest neighbor
   c) Naïve Bayes
3) Feature selection
4) Classifying documents into functional categories
5) Gene annotation

methods, and assigns appropriate functions to articles with an accuracy of 72%. The maximum entropy method provides confidence measures that correlate well with performance. Once function codes are assigned to abstracts, a voting scheme can be used to combine the assigned codes from multiple abstracts relevant to a single gene into an annotation for that gene. We thus show that statistical text analysis methods are able to automatically access relevant biological concepts from text and that information can be further used to annotate genes.

8.1 Functional vocabularies and gene annotation

In order to provide some standards for describing gene function, investigators have developed controlled vocabularies for annotation. The vocabularies include a pioneering classification for *Escherichia coli* gene function (Riley 1993), the Munich Information Center for Protein Sequences (MIPS) classification (Mewes, Frishman et al. 2000), and Gene Ontology (GO) Consortium's recent widespread effort across multiple organisms (Ashburner, Ball et al. 2000). These vocabularies contain a set of codes associated with specific genetic attributes and functions. They provide a distilled set of biological concepts that can be used to make exact statements in biology. These vocabularies are important in text mining because they provide a basic set of concrete definitions.

Great effort has been made in the last decade to begin assigning function codes to genes, to facilitate large-scale analysis. The goal is to use high quality evidence to create reliable information resources about genes. This is currently achieved by manual assignment or annotation of specific codes to genes. The value of such resources in bioinformatics is tremendous. They can be used as a starting point for analysis of data collected on thousands of genes. A large knowledge resource covering many functions and many genes offers the possibility of very comprehensive analysis. For example, an investigator can infer the function of an uncharacterized gene by obtaining and comparing primary data from it to primary data from other annotated genes. Furthermore, annotated genes can be used as gold standards in testing bioinformatics algorithms.

Unfortunately, annotating genes with these controlled vocabulary codes is a labor-intensive task. An expert inspects the literature (and, in principle, other available data) associated with each gene to determine the appropriate function code. It is likely that one-time annotation will not be sufficient; as our knowledge of biology increases and expands into new areas, the vocabularies will undergo refinement and coding may need to be repeated. Therefore, tools to automate this process are critical to building such large knowledge resources.

In the next section we briefly outline some well-known functional vocabularies.

8.1.1 Gene Ontology

Gene Ontology (GO) is one of the most widely used functional ontologies. We introduced it in Chapter 1, and have been using it as a gold standard to validate our algorithms in other chapters. Gene Ontology is a functional vocabulary that was built for general application to all organisms. It is regularly revised and updated as more and more genes from different organisms are being annotated with it. Beyond being simply a list of terms, Gene Ontology is a hierarchically arranged set of codes. Broad terms are at the top of the hierarchy, and more specific terms are at the bottom. When a gene is assigned a specific term, the more general parent terms of that term are implied functional assignments. For example *oxidoreductase activity* (GO:0016491) is a specific type of *catalytic activity* (GO:0003824), which in turn is a specific type of *molecular function* (GO:0003674). So a gene assigned the function *oxidoreductase activity*, carries with it implicit assignments of those other two more general terms (see Figure 8.1). However, a gene assigned the *catalytic activity* function term does not imply that the gene necessarily has *oxidoreductase activity* as a function.

The advantage of the hierarchical approach is twofold. First genes can be annotated as specifically as possible, given the information available. As more information becomes available about the genes, the annotations can be refined and made more specific. Second, the hierarchical structure facilitates general as well as specific queries. So if a user wishes to obtain all of the genes that have *catalytic activity*, any gene assigned that term, or any of its more specific descendent terms, will be obtained.

Gene Ontology is organized into three broad components: molecular function, cellular compartment, and biological process (see Figure 8.1). Molecular function terms describe the biochemical reactions that the gene's protein product is directly involved in. For example, the enzymatic reaction

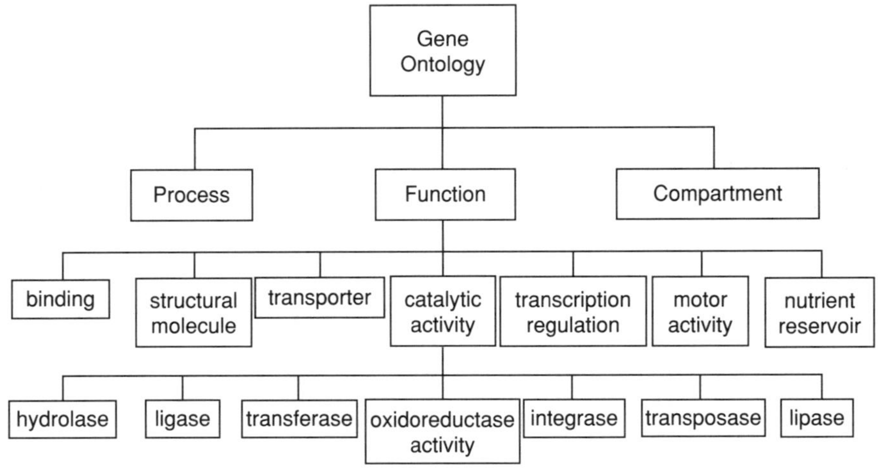

Figure 8.1 *Gene Ontology schematic.* Gene ontology (GO) has three major branches: *biological process*, *molecular function*, and *cellular compartment*. To demonstrate the hierarchical nature of GO we list selected descendents of *molecular function*, and selected descendents of *catalytic activity.* Any gene assigned a descendent of *catalytic activity*, such as *oxidoreductase activity*, is also assigned the broader functions *catalytic activity* and *molecular function* implicitly.

that a protein catalyzes can be considered its molecular function. Cellular compartment terms describe specific locations in the cell; these terms are used to describe the subcellular location of the gene's protein product. For example, the protein complex that a gene's protein product is a part of can be considered its cellular location. Biological process is a set of broad terms that describe the biological role that each gene plays. These terms include terms ranging from processes that occur at the organism level, such as organ development, to the most basic cellular processes, such as metabolism.

Gene Ontology codes are assigned to genes by many different means. After a professional assigns a term to a gene, she assigns an evidence code indicating the source of the information that suggested the term assignment. There are 12 different evidence codes (see Table 8.1). The most reliable annotations are taken from the scientific literature—these are coded as "Traceable Author Statements". Other high quality annotations are obtained from direct experimental data. These quality annotations are often the most time consuming to obtain as well, as they require detailed examination of the scientific literature. Much less reliable are the "Inferred from Sequence Similarity" annotations or "Inferred from Reviewed Computational Analysis" annotations that have been made on the basis of sequence similarity to another already annotated gene or computational analysis. Even less reliable than those are the "Inferred from Electronic Annotation" annotations; these annotations have been transferred from external databases or automatically by sequence similarity searches that have not been reviewed by any curator.

Table 8.1 *Evidence codes for Gene Ontology.* The different evidence codes used to identify the source for Gene Ontology code assignments to genes. This information was obtained from the Gene Ontology web site.

Evidence code	Abbreviation	Notes
Traceable author statement	TAS	
Inferred by curator	IC	
Inferred from direct assay	IDA	– Enzyme assays – In vitro reconstitution (e.g. transcription) – Immunofluorescence (for cellular component) – Cell fractionation (for cellular component) – Physical interaction/binding
Inferred from mutant phenotype	IMP	– Any gene mutation/knockout – Overexpression/ectopic expression of wild-type or mutant genes – Anti-sense experiments – RNAi experiments – Specific protein inhibitors
Inferred from genetic interaction	IGI	– "Traditional" genetic interactions such as suppressors, synthetic lethals, etc. –Functional complementation – Rescue experiments – Inference about one gene drawn from the phenotype of a mutation in a different gene.
Inferred from physical interaction	IPI	– 2-hybrid interactions – Co-purification – Co-immunoprecipitation – Ion/protein binding experiments
Inferred from expression pattern	IEP	– Transcript levels (e.g. Northerns, microarray data) – Protein levels (e.g. Western blots)
Inferred from sequence or structural similarity	ISS	– Sequence similarity (homologue of/most closely related to) – Recognized domains – Structural similarity – Southern blotting

Continued

Table 8.1 *Continued*

Evidence code	Abbreviation	Notes
Inferred from reviewed computational analysis	RCA	– Large-scale protein–protein interaction experiments – Microarray experiments – Integration of large-scale datasets of several types – Text-based computation
Inferred from electronic annotation	IEA	
Non-traceable author statement	NAS	
No biological data available	ND	

8.1.2 Enzyme Commission

One of the oldest and well-known functional vocabularies is the Enzyme Commission (EC) classification scheme. This vocabulary only addresses the definition and classification of enzymes by the molecular function they catalyze. In practice each EC number is written as four numbers with periods in between. The left-most number defines the broadest classification of enzymes and can range from 1 to 6. In Table 8.2 we list these broad classifications. The numbers to the right define the enzymatic reaction more specifically. For example, consider an enzyme annotated with the EC number 3.2.1.4. The number 3 specifies that the enzyme is a hydrolase. The number 2 specifies that it is a hydrolase that breaks glycosidic bonds. The number 1 specifies that the bonds that are hydrolyzed by these enzymes are actually O-glycosyl bonds. And the final number 4 indicates that the enzyme catalyzes the hydrolysis of 1,4-beta-D-glucosidic linkages in cellulose, lichenin and cereal beta-D-glucans. Like Gene Ontology, the EC classification is a hierarchical vocabulary, but with only four levels. Many other biological vocabularies have been constructed around the EC classification.

8.1.3 Kyoto Encyclopedia of Genes and Genomes

Another ontology is the Kyoto Encyclopedia of Genes and Genomes (KEGG) (Kanehisa, Goto et al. 2004). This ontology seeks to unify functional

Table 8.2 *Enzyme Commission (EC) classification categories*. Listed below are the broad EC categories. They correspond to the first of the four numbers that define each enzyme classification.

Class 1	Oxidoreductase
Class 2	Transferase
Class 3	Hydrolase
Class 4	Lyase
Class 5	Isomerase
Class 6	Ligase

information about genes from over 100 organisms. In addition, KEGG seeks to provide a unified resource that characterizes the chemical reactions that each of the genes are involved in and also catalogs protein–protein interactions and molecular pathways. It contains three separate categories of structured information: structured information about genes, protein networks, and chemical intermediates.

Gene information is stored in the GENES database. This database contains a listing of over half a million genes from more than 150 sequenced organisms. These genes are annotated with KO (Kegg Orthology) numbers that correspond to specific biological functions. The idea is that genes from different organisms with the same KO number have the same biological function and also are evolutionarily related. The KO numbers are an extension and modification of the Enzyme Classification (EC) numbers described above. There are five broad categories of pathway functions used in the KEGG ontology: metabolism, genetic information processing, environmental information processing, cellular processes, and human disease. There are 24 subcategories at the next level. The third level is individual pathway maps. The fourth level is specific KO numbers that correspond to orthologous groups of genes sharing the same biological function. The actual orthologous groups of genes are assembled semi-automatically.

KEGG also contains a large repository of structured information about chemical intermediates and metabolites. It contains entries for over 10,000 chemical intermediates and more than 10,000 carbohydrate entries as well. A database of reactions contains over 5,000 reactions; each entry lists the products and reactants for these reactions.

The information on molecular interactions between proteins and pathways is contained in the PATHWAYS database. This database contains manually compiled networks of functional significance, including metabolic pathways and protein–protein interactions. We will discuss learning genetic networks and pathways more extensively in Chapter 10.

8.2 Text classification

Successful recognition of the biological function described in a short segment of text is critical to gene function annotation. Functional annotation with text requires us to determine the most relevant segments of text, and to recognize its meaning. Text classification methods can help us focus in on the critical pieces of text and they can also help us decide what function is being described in that piece of text.

Given a piece of text, the goal is to classify it into the most appropriate biological function. If the text can be classified with confidence, it is likely a relevant segment of text. These techniques fall under the heading of supervised machine learning (Manning and Schutze 1999). An alternate use of text classification is to use it to separate the relevant from irrelevant segments of text, so that they can be subject to further manual scrutiny.

Classification algorithms require examples of text that have already been classified. For example, a large collection of text segments that are associated with known biological function would suffice. These examples of text might be PubMed abstracts or specific selected sentences from scientific articles. These examples are known as "training" examples. Ideally, experts should select the training examples carefully. After the classifier is trained on these examples, it can classify unseen "test" cases. In our case, the test cases are segments of text relevant to a particular gene of unknown function.

These methods are similar to some of the methods we described for gene expression profile classification in Chapter 2. However, text classification is more challenging than gene expression profile classification. For one thing the set of features is much larger. For example, documents might be classified based on the words that are contained or not contained in it. In that case, the features examined are the vocabulary words. The number of vocabulary words that are important to our analysis is often enormous. So instead of looking at 10–100 gene expression measurements to classify a gene, we look at the presence or lack of tens of thousands of specific words to classify a document. Since the feature set is so large, the available data are comparatively inadequate. So complex statistical models for classification work poorly for document classification. Sophisticated Bayesian networks or neural networks that require many parameters to be fit are probably more effective in the world of gene expression classification. The relative paucity of training examples compared to the number of features results in degenerate fitting. In document classification the features must be explained by either fewer parameters or by very constrained parameters. Simple classification methods that make strong assumptions about the data can be more effective for text classification.

Here we present several examples of well-known text classification algorithms. Naive Bayes and nearest neighbor classification schemes are easy-to-understand and effective methods. Maximum entropy modeling is effective in generic text classification tasks, and has gained recent popularity (Ratnaparkhi 1997; Nigam, Lafferty et al. 1999).

Besides classifying text, it is also important to know how confident the classifier is about its classification. For one thing, we would only want to focus our attention on high confidence statements in the text that are certainly describing biological function. Low confidence predictions may lead us astray, and cause erroneous predictions to be made about a gene's function. In addition, confidence scores provide a scheme to combine predictions from different pieces of text.

Text classification can be widely applied to many different areas outside gene annotation as well. For example, text classification can be used to identify the key sentences that describe protein–protein interactions, and help to elucidate networks of interactions between different proteins; we will explore this further in Chapter 10. Or they can be used to scan the literature for papers in which a specific sort of assay was attempted, or to identify papers that are relevant to a particular organism. They offer an opportunity to scan and understand a large body of literature at a high level.

8.3 Nearest neighbor classification

Nearest neighbor classification can be applied to document classification much in the same way that it can be applied to gene expression profiles. In nearest neighbor classification, a distance metric is employed to calculate the distance between the word vector of an unclassified document in the test set and each of the abstracts in the training set. Documents can be first converted into weighted word vectors as described in Chapter 3. A cosine-based distance metric between the word vectors can be used to assess document similarity:

$$\text{dist}(a,b) = 1 - \frac{a \cdot b}{\|a\| \, \|b\|}$$

where a and b are vectors of word counts. Alternatively a and b can be document vectors in latent semantic indexing space as described in Chapter 3. Typically given a test document, the k most similar documents in the training set are obtained. The classifications of each of the k documents are noted, and the document is assigned the classification that corresponds to the greatest number of those documents.

The basic assumption behind nearest neighbor classification is that two documents that use the same sort of words in general are likely to have the same meaning. This method is of course very sensitive to the selection of the document weighting scheme and the choice of document metric. Not only are a wide variety of metrics available, but words can also be weighted differentially depending on their preconceived value to the classification task at hand.

8.4 Naive Bayes classification

Another simple and effective document classification strategy is the naive Bayes classifier (Manning and Schutze 1999). There are two basic assumptions behind naive Bayes. First, is that the probability that a word is found in a document is a function of the classification of the document, and that different classes of documents have different distributions of words. Second, naive Bayes assumes that the words within the document are independently distributed. That is, the presence of one word has no bearing on whether another word will be present or not. In reality this is clearly not the case. For example a document that has the word “transduction” in it will very likely have the word “signal” in it as well if it is a paper about signal transduction. So this is a strong and inaccurate assumption in principle, but in practice naive Bayes is often effective. The large number of vocabulary words compared to the relatively few documents available requires effective classification techniques to make strong assumptions about the data.

Therefore, given a document, and the class that it belongs to, the probability of the document is the multiplicative product of the probability of each of the individual words occurring in that class. So the probability of a document d given that it is a member of a class c can be calculated:

$$P(d|c) = \prod_{w \in d} P(w|c)$$

In naive Bayes, the probability of each word appearing in a document of a certain class is estimated directly from the training data. Given a large collection of documents that pertain to a specific class, we estimate the probability to be the number of documents that the word occurs in, divided by the total number of documents. There is the concern that a specific word may show up in the training documents too rarely or not at all. Under these circumstances the probability of a document containing that word will be zero. To avoid this situation we employ pseudo-counts of words that are based on a background distribution:

$$P(w_i|c) = \frac{s_i + \sum_{d \in C} I(w_i, d)}{s + N_{d \in C}}$$

In this equation, c is a specific class, d is a training document within the class c, and $N_{d \in C}$ is the number of documents within the class c. $I(w, d)$ is an indicator function that is 1 if the word w is in the document d, 0 otherwise. The term s_i represents the number of times we would expect to see word i in s documents. One way of calculating s_i is to calculate a background distribution from all of the documents available and to multiply it by s. These terms are added in as pseudo-counts to compensate for unseen events and avoid zero probabilities. If very few documents are available, then the pseudo-counts determine the probability. If large numbers of documents and words are available, the influence of the pseudo-counts is minimal.

Given an unclassified document, we can use Bayes theorem to calculate the probability that a class c is the correct classification of a document d. Under the naive Bayes model:

$$P(c|d) = \frac{P(c)P(d|c)}{P(d)} \sim P(c) \prod_{w_i \in d} P(w_i|d)$$

Here $P(c)$ represents the prior probability of class c. That is, it is the background distribution of class c; it can be estimated as the fraction of documents in the training set that correspond to that classification. $P(d)$ is the probability of a document; this is a constant term across the different classes.

To predict the class of an unseen test set document, we calculate the above probability of the document given each class and multiply it by the prior probability of that class. The class that receives the highest value is predicted.

While naive Bayes is a very easy method to understand and implement, its performance is usually comparable to the best classifiers. The limitation is in the strong independence assumption that is made. While that assumption allows the construction of an easy to fit probabilistic framework, it also leads to a classifier that over-commits to classifications. The presence of multiple key words in a document that imply a specific function leads to unreasonably high probabilities since the presence of these keywords was assumed to be independent. The result is a classifier that gives extreme probabilities in the case that the classification is obvious, but deals with ambiguous cases poorly.

8.5 Maximum entropy classification

Maximum entropy modeling is a classification method that has proven to be effective in many text classification tasks (Ratnaparkhi 1997; Nigam,

Lafferty et al. 1999). It is similar to naive Bayes classifications in terms of the probabilistic framework to classify documents. But the framework is based on different assumptions. Entropy can be used to characterize probabilistic models used for classification. Low entropy models depend on making many distinctions when classifying documents, and can suffer from over-interpretation of the training data. High entropy models make fewer distinctions but do not take full advantage of the signal within the training data. For example, the uniform distribution has the greatest possible entropy; under this distribution all observations are equally likely.

Maximum entropy methods are based on the strong assumption that the best models are those with the highest entropy that are still consistent with the training data. The parameters are selected via an iterative algorithm to insure that the distribution has the highest possible entropy. This strategy biases us to pick a probabilistic distribution for classification that is least prejudiced by the observed data in the training set. The result is a classifier that does not over-fit the training data, and make spurious classifications based on limited information. One advantage of maximum entropy classification is that, in addition to assigning a classification, it provides a probability of each assignment being correct.

In maximum entropy classification, the user defines category specific "features". Each feature $f_i(d,c)$ is a binary function of any document d and any class (or code) c. In this application each feature f_i is defined relative to a specific word w_i and class c_i. The feature $f_i(d,c)$ is unity only if d contains w_i and c is c_i. For example, one feature f_{example} might be

$$f_{\text{example}}(d,\ c) = \begin{cases} 1 & \text{if "cell"} \in d;\ c = \text{` metabolism '} \\ 0 & \text{otherwise} \end{cases} \tag{8.1}$$

where w_{example} is "cell" and c_{example} is "metabolism".

The goal in maximum entropy is to define a probabilistic distribution that assigns a probability to each possible document and class. The entropy of the distribution, P, can be calculated as follows:

$$H(P) = -\sum_j \sum_i P(d_j,c_i) \log\left(P(d_j,c_i)\right)$$

where c is a class, d is a document.

To help demonstrate we present a simple example of documents that can be classified into two different classes, + and −. We define two features for each classification that are based on the presence of a specific word in the document. So we have four features altogether f_{1+}, f_{2+}, f_{1-}, and f_{2-}. So given a document and a class there is a possible total of seven feature vectors

Table 8.3 *Maximum entropy example.* Here we give a simple example of the maximum entropy modeling formulation. We define two features for each of the two classes. There are seven possible feature vectors that can be defined over these features. Every document can, given the class that it belongs to, be mapped to one of these seven vectors. The highest entropy probability distribution is to assign equal weight to each vector; that distribution has 2.81 bits of entropy. The lowest possible entropy distribution is to assign all of the probability mass to a single vector; that distribution has 0 bits of entropy. Now suppose that we wanted the probability distribution to be such that the expectation of each feature was a specific value. That constraint greatly limits our choices of distributions. We provide two possible distributions. The lower entropy distribution has zero probability assigned to three of the vectors and has 1.74 bits of entropy. The higher entropy distribution has non-zero values for all of the vectors, and is closer to a uniform distribution. It has 2.55 bits of entropy. It is a more plausible distribution.

	+ class		− class		Highest entropy distribution	Lowest entropy distribution	Lower entropy distribution	Higher entropy distribution
Feature	f_{1+}	f_{2+}	f_{1-}	f_{2-}				
vec 1	1	1	0	0	0.143	1	0.05	0.20
vec 2	0	1	0	0	0.143	0	0.35	0.20
vec 3	1	0	0	0	0.143	0	0.20	0.05
vec 4	0	0	0	0	0.143	0	0	0.05
vec 5	0	0	1	1	0.143	0	0.40	0.30
vec 6	0	0	0	1	0.143	0	0	0.10
vec 7	0	0	1	0	0.143	0	0	0.10
expectation	0.25	0.40	0.40	0.40				
Entropy					2.81	0	1.74	2.55

that can be defined (see Table 8.3). Our goal is to define probabilities over these seven possible feature vectors. If we were completely unconstrained, the highest entropy probability distribution possible would be the uniform distribution; each case is assigned equal probability. However, this does not provide a very helpful statistical model. Now suppose that we had some constraints. Say, we wanted to define a distribution such that f_{1+} had an expectation of 0.25, while the other features had an expectation of 0.4. In other words we wish to define a distribution among those seven possibilities such that f_{1+} is set to 1 in 0.25 of the cases and 0 in 0.75 of the cases. In Table 8.3 we present a high and low entropy possibility that conforms to this constraint. Notice the higher entropy scenario has fewer extreme probabilities, and is closer to the uniform distribution. We assert that this high entropy distribution is closer to the "real" distribution.

Our goal is to identify the distribution P that maximizes the entropy while also resembling the training data. We apply a simple requirement on the choice of distribution. We constrain our choice of P so that the fraction of documents that a feature is observed in a data set is equal to the expectation of that feature under the model *P*. That is, given a training data set *D* with real documents and class assignments, we can calculate the fraction of documents that each feature is observed in:

$$g = \frac{1}{|D|} \sum_{d \in D} f(d, c_d)$$

where *d* is a document in the training set, c_d is the document's true classification and $|D|$ is the number of documents in the training set. The expectation of the feature is the average fraction of times we would expect the feature to occur in a random set of data. The expectation of the same feature under a distribution *P* can be calculated:

$$\bar{g} = \sum_i \sum_j P(d_j, c_i) f(d_j, c_i)$$

Notice, in theory this is calculated over all possible documents *d* and not just documents in the training set. It is quite reasonable to expect that the fraction of times a feature is observed in practice and the expected fraction of times a feature is observed under a model should be the same if the model approximates reality well. We thus require the selection of a distribution *P* such that the above two quantities for each feature are identical.

With the given feature formulation and the above constraint, it can be shown that the maximum entropy probability distribution for documents must be distributed according to an exponential distribution. So

$$P(c_j, d) = \omega \exp\left(\sum_i \lambda_i f_i(d, c_j)\right)$$

where *c* is a class, *d* is a document, the λ_i are feature weights, and ω is a constant that ensures that the probability distribution is normalized.

The probability of each class for a test document can be calculated with this exponential model:

$$P(c_j|d) = \frac{1}{Z(d)} \exp\left(\sum_i \lambda_i f_i(d, c_j)\right)$$

where *Z(d)* is a normalization constant:

$$Z(d) = \sum_c \exp\left(\sum_i \lambda_i f_i(d, c)\right)$$

Once the parameter weights are fixed, the probability that a document belongs to a particular class can be calculated with the above pair of equations. The class with the highest probability is assigned. Furthermore the probability of that class can be viewed as a measure of confidence in the prediction.

Now, the challenge is to fit the parameter weights, λ_i. Each λ_i weight is selected so that the aforementioned constraint on the probability density is satisfied: the expectation of f_i must equal its observed frequency in the training data. In principle, this expectation should be calculated over the true distribution of documents. However, in practice this distribution is unknown, so instead we estimate the expectation for the feature empirically over the training documents D. So in practice the constraints are reduced to:

$$\frac{1}{|D|} \sum_{d \in D} f_i(d, c_d) = \frac{1}{|D|} \sum_{d \in D} \sum_c P(c|d) f_i(d, c)$$

Here c_d is the correct classification of document d specified in the training set, P is the probability calculated from the statistical model, and $|D|$ is the number of documents in the training set. The left side is the observed fraction in the training data, while the right side is the expectation given a distribution P. This constraint must hold true for each of the features.

In practice, the weight parameters are fit to satisfy this constraint by iteration. One common method is generalized iterative scaling (GIS) (Ratnaparkhi 1997). In GIS, all of the weights are assigned to be zero in the first iteration. Then, in every iteration the weights are updated:

$$\lambda_i^{n+1} = \lambda_i^n + \frac{1}{C} \ln\left(\frac{g_i}{g_i^n}\right)$$

The fraction of documents each feature is observed in the data set is also computed initially – this is invariant over the iterations. Then at each iteration, the expectation for each feature is re-calculated with the given parameters. If these two terms are equal, the log of the ratio is zero and the weight is not updated. On the other hand, if the observed fraction is much larger, the log of the ratio is positive, and the weight is updated to be a larger value. If the observed fraction is smaller than the expectation the log of the ratio is negative, and the weight is updated to be a smaller value.

GIS requires that the sum of all the features for any document-class instance is always the same value C. In practice an extra non-binary feature is added to achieve this. The extra feature is assigned C minus the sum of all of the other terms. The factor C modifies the speed at which the algorithm converges and is included to insure the correctness of the algorithm. Other strategies to satisfy the constraint and find the exponential parameters are also available. The more features that are used in a classification the more complicated it is to find values of the parameters that satisfy our constraint. In this case the value C is larger, and the weights are updated slowly in each iteration so that convergence can be achieved.

8.6 Feature selection: choosing the best words for classification

In many cases there may be too many vocabulary words for quick classification. We can make intelligent decisions to include and exclude certain words from our calculation to facilitate rapid computation. In many cases, if we eliminate the right words, or features, from our analysis we can do so without costing prediction accuracy. Indeed, under many circumstances performance can actually be improved since erroneous features that can lead the analysis astray are no longer present. Feature selection is an active field in the machine learning scientific community. We will only focus on a simple, but very effective method.

Chi-squares is a statistical method that can be used to see if a certain feature occurs more or less often with a certain classification than expected. First, documents are divided by whether they have the feature and by which class they belong to. If there are only two classes, the results can be placed in a 2×2 table. See Figure 8.2 for an illustration. If the feature is a valuable one for prediction, we would expect that its presence correlates with the classification of the document.

The expected number of documents in each box can be calculated if we assume there is no relationship between the feature and the classification. We simply calculate the fraction of documents in which the feature occurs, p, and the fraction of times the classification occurs, q. The number of times we then expect a feature and a classification to occur together if they are independent is therefore Npq. Similarly the number of documents where the feature is lacking and it has a different classification can be calculated as well: $N(1-p)(1-q)$. The chi-square score can be calculated:

$$\chi^2 = \sum_i \frac{(O_i - E_i)^2}{E_i}$$

	class (+)	*class* (–)
feature (+)	A	B
feature (-)	C	D

$$P(f_+) = \frac{A+B}{A+B+C+D} \qquad P(c_+) = \frac{A+C}{A+B+C+D}$$

$$P(f_-) = \frac{C+D}{A+B+C+D} \qquad P(c_-) = \frac{B+D}{A+B+C+D}$$

$$Expected(f_+,c_+) = N \cdot P(f_+,c_+) = (A+B+C+D)\, P(f_+)P(c_+) = \frac{(A+B)(A+C)}{A+B+C+D}$$

Figure 8.2 *Chi-square testing to select features.* For each feature, all of the documents from the training data are divided into four categories: those that possess the feature and are in the class (A), those that possess the feature and are not in the class (B), those that lack the feature and are in the class (C), and finally those that lack the feature and are not in the class (D). The candidate features that we are testing are typically the presence of a specific word. We are looking for features where the class is correlated with the feature. If there is no correlation, then the probability of the feature and the class can be calculated independently with the equations below. To calculate the expected number of documents, assuming there is no relationship between the feature and the class, we simply multiply the total number of documents by the probability of the feature (or lack of the feature) and the probability of the class (or lack of the class). We calculate this value for each of the four possibilities. Comparison of these expected values assuming the feature and class are independent to the actual values in the table yields the chi-square score.

Here, O is the number of times each of the four possibilities is observed and E is the number of times each of the four possibilities is expected to occur if the feature and class are unrelated. If all of the data collected are in fact random, the chi-square score should be close to zero. The probability that it is larger than the obtained value if the data were collected randomly is given by the Chi-square distribution (the p-value) with one degree of freedom. In practice we select the words with the largest Chi-square scores and smallest p-values. These features are likely to constitute a set of features that correlate with the classifications.

This method can be easily extended to multiple classifications. Instead of a 2×2 table, a $2 \times n$ table needs to be created. This time for each classification c_i, the fraction of times that classification occurs can be calculated, q_i. Expected values of each box can be calculated, and a chi-square score can be calculated in the same way as above. The only difference is that instead of summing over four boxes, we sum over $2n$ boxes. The statistical significance can be determined by looking to the chi-square distribution with $n - 1$ degrees of freedom.

8.7 Classifying documents into functional categories

To demonstrate the effectiveness of using text classifiers to discern biological function in text, we train a maximum entropy classifier to recognize the text most relevant to Gene Ontology functions. Later in this chapter we will show how such a classifier can be used to annotate a gene by combining the GO code classifications from all of its abstracts.

Here we (1) evaluate the performance of a maximum entropy document classifier to obtain genetic functions and (2) annotate the gene based on the literature. To evaluate document classification, we use *accuracy*; to evaluate gene annotation we use *precision* and *recall*. Accuracy is the percentage of predictions on a document test set for which the classifier prediction was correct. Precision is the percentage of annotated genes that are true positives. Recall is the percentage of genes that truly have the function that are true positives.

We conduct experiments to annotate *Saccharomyces cerevisiae* (yeast) genes with codes from a subset of GO. We choose this organism because many of its genes have manually curated GO annotations that can be used as a gold standard. We used a high quality list of PubMed citations hand assigned to relevant genes by the curators of the Saccharomyces Genome Database (SGD) (Cherry, Adler et al. 1998; Ball, Dolinski et al. 2000).

Since the crux of our annotation strategy is a document classifier, we compare it to the two other types of classifiers described. After establishing the effectiveness of a maximum entropy classifier, we evaluate using the classifier's probabilistic estimates as robust confidence estimates of prediction. Finally we introduce a voting scheme to combine document classifications into gene annotations. Since our classifications of documents have reliable confidence estimates, our annotations of genes should also. At higher confidence cutoff values, better precision is achieved since the predictions are more certain, but at the cost of lower recall since low confidence correct annotations are missed.

In total we scrutinize 21 gene function categories (see Table 8.4a). All codes are biological process codes that are relevant to yeast. The unclassified text is assigned to categories based on similarities with the training examples.

8.8 Comparing classifiers

We compare the classification accuracy of two different classifier families, naive Bayes and nearest neighbor, to maximum entropy classification. We tested the different document classifiers described above to predict the subject matter of the documents in two test sets, *test2000* and *test2001*.

We begin by creating a corpus containing at least 1000 PubMed abstracts relevant to each functional category. These comprise the *training* sets for the classification algorithms. The best approach to this problem would be careful examination of many articles by qualified experts. However, obtaining a large volume of abstracts in this manner is very difficult. Instead, we use MeSH term headings and title words to query PubMed for relevant abstracts as a surrogate(Bachrach and Charen 1978). MeSH headings are keywords that have been carefully assigned to documents by curartors that have studied them closely. For each code we identify the most relevant MeSH terms that were semantically similar to the code or one of its children in GO. Then, we use those terms to construct a PubMed query for each GO code; most queries included specific MeSH terms as a major heading for the article. For many categories, such as signal transduction, an obvious MeSH term is available; other categories require use of an appropriate combination of title words and MeSH terms. The queries also included the "genes" or "genetics" MeSH headings to insure that the article is biologically focused. To balance the sizes of the training sets, we adjusted the publication date so that approximately 1000 abstracts could be obtained. We construct training and test corpora of documents for the 21 GO codes by searching PubMed with those queries.

The queries and the number of abstracts per GO code are listed in Table 8.4(a). We split the results into three sets based on publication date; documents published before 2000 constitute the *training* set, documents published in 2000 constitute the *test2000* set, and documents published in 2001 constitute the *test2001* set. A few of the documents are relevant to more than one GO code (see Table 8.4b).

We use the abstracts and title fields from the PubMed records for the *training* set only. Since the titles were sometimes used to select the articles, we omit the title from the document when testing. From these documents, we find 63,992 unique tokens by tokenizing on white space, punctuation,

Table 8.4 *The training and testing corpus.* (a) This table lists the gene ontology code in the first column and the PubMed query used to obtain abstracts in the final column. For the *training* data set, the articles were obtained by using the query as listed in the table. The *test2000* and *test 2001* data sets were obtained by modification of the publication date limit to restrict articles to those published in 2000 and 2001, respectively. Titles were omitted from the test data sets. The table also lists the number of articles obtained for each category for the training and test sets. (b) Some of the articles within the training set were obtained in more than a single query; thus these articles have multiple GO classifications. This table lists the number of abstracts in each data set and the number of abstracts with 1, 2, 3, and 4 relevant codes.

(a)

GO code	Training	Test2000	Test2001	PubMed query
metabolism	1005	225	30	"(metabolism[MAJR]) AND Genes[MH] AND 1989:1999[DP]"
cell cycle	1085	303	19	"(cell cycle[MAJR]) AND Genes[MH] AND 1996:1999[DP]"
signal transduction	1168	302	25	"(signal transduction[MAJR]) AND Genes[MH] AND 1995:1999[DP]"
oncogenesis	1043	168	15	"(cell transformation, neoplastic[MAJR]) AND Genes[MH] AND 1994:1999[DP]"
cell death	1154	434	28	"(cell death[MAJR]) AND Genes[MH] AND 1997:1999[DP]"
meiosis	1003	151	7	"((meiosis[MAJR])) AND (Genes[MH] OR Proteins[MH]) AND 1986:1999[DP]"
intracellular protein traffic	1107	322	28	"(endocytosis[MAJR] OR exocytosis[MAJR] OR transport vesicles[MAJR] OR protein transport[MAJR] OR nucleocytoplasmic[TI]) AND (Genetics[MH]) AND 1994:1999[DP]"
cell adhesion	1025	133	5	"(cell adhesion[MAJR]) AND (genetics[MH]) AND 1993:1999[DP]"
cell motility	1094	269	23	"(cell movement[MAJR]) AND (Genetics[MH]) AND 1995:1999[DP]"
sporulation	847	49	0	"(sporulation[TI]) AND (genetics[MH]) AND 1940:1999[DP]"
membrane fusion	317	58	4	"(membrane fusion[MAJR]) AND (Genetics[MH]) AND 1940:1999[DP]"

autophagy	177	22	1	"(autophagy[TI] OR autophagocytosis[MAJR]) AND (Proteins[MH] OR Genes[MH]) AND 1940:1999[DP]"
cell fusion	740	20	0	"(cell fusion[MAJR] OR (mating[TI] AND Saccharomyces Cerevisiae[MAJR]) AND (Genetics[MH]) AND 1940:1999[DP]"
stress response	1068	253	22	"(Wounds[MAJR] OR DNA repair[MAJR] OR DNA Damage[MAJR] OR Heat-Shock Response[MAJR] OR stress[MAJR] OR starvation[TI] OR soxR[TI] OR (oxidation-reduction[MAJR] NOT Electron-Transport[MAJR])) AND (Genes[MH]) AND 1996:1999[DP]"
cell proliferation	394	0	0	"(cell proliferation[TI]) AND (Genes[MH]) AND 1940:1999[DP]"
biogenesis	1023	132	4	"(biogenesis[TI] OR ((cell wall[MAJR] OR cell membrane structures[MAJR] OR cytoplasmic structures[MAJR]) AND (organization[TI] OR arrangement[TI]))) AND (Genetics[MH]) AND 1984:1999[DP]"
cell-cell signalling	237	41	0	"(synaptic transmission[MAJR] OR synapses[MAJR] OR gap junctions[MAJR]) AND (Genes[MH]) AND
invasive growth	492	52	4	"((invasive[TI] AND growth[TI]) OR neoplasm invasiveness[MAJR]) AND (Genetics[MH]) AND 1940:1999[DP]"
transport	1022	84	8	"(biological transport[MAJR] OR transport[TI]) AND (Genes[MH]) AND 1985:1999[DP]"

Continued

Table 8.4 *Continued*

GO code	Training	Test2000	Test2001	PubMed query
ion homeostasis	424	64	5	"((na[TI] OR k[TI] OR ion[TI] OR calcium[TI] OR sodium[TI] OR hydrogen[TI] OR potassium[TI] OR pH[TI] OR water[TI])AND (concentration[TI] OR senses[TI] OR sensing[TI] OR homeostasis[TI] OR homeostasis[MAJR]) AND (genetics[MH]) AND 1940:1999[DP]"
chemi-mechanical coupling	1011	147	6	"(contractile proteins[MAJR] OR kinesins[MAJR]) AND (Genes[MH]) AND 1993:1999[DP]"

(b)

	# Articles with *N* codes				
Corpus	1	2	3	4	Total articles
training	15444	888	60	9	16401
test2000	2682	231	27	1	2941
test2001	184	22	2	0	208

and common non-alphanumeric characters such as hyphens and parentheses. From these, we exclude stopwords, which were defined as tokens that appeared in four or less or 10,000 or more documents. This leaves a total of 15,741 unique words. Then, we represent documents as 15,741 dimensional vectors of word counts.

We train each classifier on the *training* set and fit their parameters by maximizing performance on the *test2000* data set. The results of the classification trials on the *test2000* data set are summarized in Table 8.5(a).

The parameters we experiment with include different vocabulary sizes. For the naive Bayes and nearest neighbor classifiers, we use a chi-square method to identify the words whose distribution is most skewed across all 21 GO codes in the training set; see Section 8.6. We take only the words with the highest scoring chi-square values as features. We experiment with using different vocabulary sizes including the full 15,741 words, and also with reduced vocabularies of 100, 500, 1000, and 5000 words selected by chi-square score. For nearest neighbor classification we experiment with

Table 8.5 *Classification performance of different supervised machine learning algorithms.* (a) Classification performance for algorithms on the test2000 data set across different parameters. For maximum entropy classification we attempted different numbers of word-features/code; also we tested the accuracy at each iteration of the GIS optimization algorithm. We report in each column the number of words/code, the highest accuracy obtained, and the first iteration obtaining that highest accuracy. For naive Bayes classification, we calculated accuracy on different vocabularies. The size of the vocabulary and the accuracy is reported in each column. For nearest neighbor classification we calculated accuracy for different numbers of neighbors and different vocabularies. The accuracy data is reported in a grid, with different numbers of neighbors for each row, and with different vocabularies for each column. (b) For each classification algorithm we fix the optimal parameters based on the data in (a). The classifier is run with optimal parameters on test2001; the accuracy is reported in this table.

(a)

Maximum entropy

# words / code	10	50	100	250	500	750	1000	2000	4000
Iteration	83	109	186	104	169	104	199	65	59
Accuracy	68.62	72.73	72.8	72.56	72.83	71.54	71.44	69.47	67.66

Naive Bayes

# words	100	500	1000	5000	All
Accuracy	63.89	66.92	66.88	65.59	63.79

Nearest neighbor

	# words				
neighbors	100	500	1000	5000	All
1	58.04	54.06	52.84	53.28	52.19
5	60.52	57.53	57.84	58.38	56.82
20	59.71	59.91	60.8	61.88	61.24
50	59.23	60.39	61.85	62.9	62.26

(b)

Classifier	Accuracy
Maximum entropy (100 words/category)	72.12
Naive Bayes (500 words)	59.62
Nearest neighbor (5000 words, 50 neighbors)	61.54

different numbers of neighbors as well. Because of its formulation, we use a slightly different strategy to vary the vocabulary size for the maximum entropy classifier. Each feature is defined relative to a code and word; the feature is assigned one for a document only if that word is in the document and the document is relevant to that code. We use the chi-square test to find the words that are most unevenly distributed when comparing abstracts relevant to a given code to all other abstracts. A word that scores high against a code is used with that code to define a feature. We take only the words with the highest scoring chi-square values for each code. We experiment with different numbers of features per code; we attempt a total of 210 (10/code), 1050 (50/code), 2100 (100/code), 5250 (250/code), 10,500 (500/code), 15,750 (750/code), 21,000 (1000/code), 42,000 (2000/code), and 84,000 (4000/code) features.

For maximum entropy classification trials we report the highest accuracy over the 200 generalized iterative scaling (GIS) iterations for different vocabulary sizes. Based on these results we choose 100 words/code as the optimal feature size for maximum entropy classification. While 500 words/code performs slightly better, it is less robust than 100 words. Either doubling to 200 words or splitting to 50 words does not significantly affect performance; however going from 500 to 750 words degrades the performance on the *test2000* set by more than a full percent. Naive Bayes performs best with a vocabulary of 500 words; nearest neighbor performs best with 50 neighbors and 5000 words. Table 8.5(b) lists the performance of each of the classifiers on the smaller *test2001* data set after parameter optimization on the *test2000* data set. Maximum entropy has the best performance (72.12% accuracy) compared to nearest neighbor (61.54%) and naive Bayes (59.62%). Results of maximum entropy classification for individual categories are reported in Table 8.6.

Our findings about the effectiveness of maximum entropy classification performance are consistent with recent reports within the statistical natural language processing literature (Nigam, Lafferty et al. 1999; Rosenfeld 2000). Frequently in statistical natural language modeling tasks, there is insufficient data to adequately estimate the large number of parameters involved. Naive Bayes compensates for this limitation by making a strong independence assumption that the words are associated with codes independent of each other. This is untrue in text classification tasks, where many dependencies exist between words. Maximum entropy relaxes this assumption, by allowing differential weights for different word-code associations.

It should be recognized that the classification of documents is not exact; there are often ambiguities. Funk and Reid examined 760 biomedical articles that had been assigned MeSH headings by two experts (Funk and Reid 1983). They found that the major MeSH headings, controlled vocabulary terms that represent the central concepts of the document, were assigned with only a 61.1% consistency. This study illustrates the subjective nature of document classification; the same sort of inconsistency may fundamentally limit performance on documents analyzed in our study.

While the document classifier may misclassify a document, the correct class is almost always assigned a high probability and is contained in the top

Table 8.6 *Classification accuracy for different categories.* For each code listed in the first column we list the number of articles in the *test2000* set for which that code is relevant in the second column. The "Exact match" column lists the percentage of articles for which the classifier predicts the code listed. Since some abstracts have multiple correct codes, the "Partial match" column lists the percentage of articles for which the classifier assigned any correct code to the article, even if it is not the listed code.

Category	Number	Exact match	Partial match
metabolism	225	67.56%	74.22%
cell cycle	303	45.87%	68.65%
signal transduction	302	59.93%	67.55%
oncogenesis	168	63.10%	70.83%
cell death	434	75.81%	79.72%
meiosis	151	77.48%	82.78%
Intracellular protein traffic	322	68.63%	72.67%
cell adhesion	133	66.17%	70.68%
cell motility	269	71.38%	74.35%
sporulation	49	73.47%	81.63%
membrane fusion	58	48.28%	53.45%
autophagy	22	59.09%	68.18%
cell fusion	20	65.00%	75.00%
stress response	253	64.82%	73.52%
cell proliferation	0	-	-
biogenesis	132	58.33%	61.36%
cell-cell signalling	41	73.17%	92.68%
invasive growth	52	69.23%	71.15%
transport	84	60.71%	70.24%
ion homeostasis	64	79.69%	81.25%
chemi-mechanical coupling	147	79.59%	82.31%

four predictions. Maximum entropy classification assigns a probability to each of the 21 codes for each abstract. A good classifier would assign the correct classification a high probability; a perfect classifier would assign the correct classification the highest probability. For abstracts in *test2000* we sorted the predicted GO codes by probabilities and calculated how often the n-th prediction was correct (Figure 8.3). The top prediction was correct 72.8% of the time as listed in Table 8.5. Predictions that were ranked greater than four rarely contained accurate predictions. The accuracy of the prediction drops off gradually with its rank.

To define a reliable voting scheme, it is critical to establish robust confidence estimates for correct document classification. We test the probability of the predicted GO code as a measure of confidence for the prediction. With reliable confidence scores, we expect that document classifications with high confidence scores are likely to have been classified correctly. To assess whether the reliability of the classifier prediction tracked with the

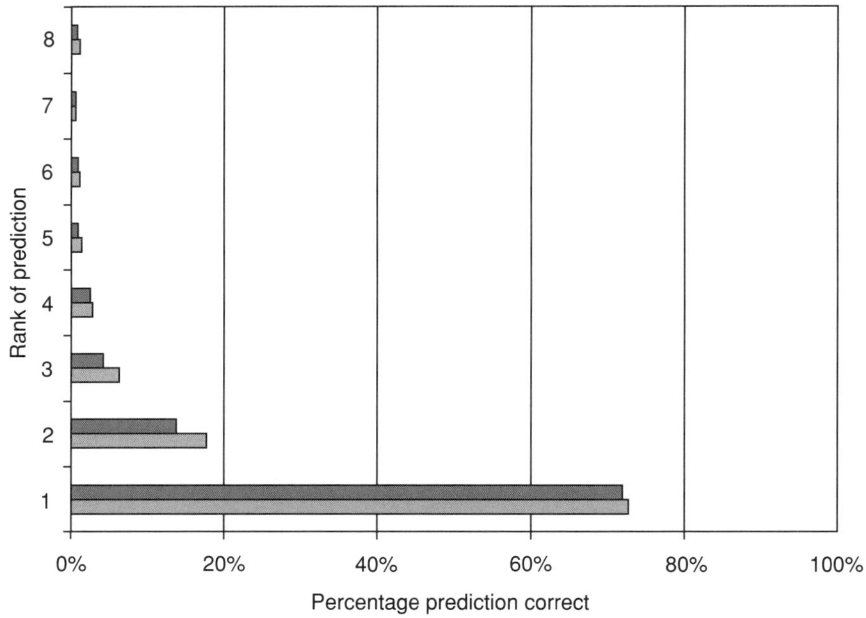

Figure 8.3 *Maximum entropy classifier ranks classifications.* The maximum entropy classifier assigns a probability to each code for the unclassified document. Here we have ranked each code by its maximum entropy assigned probability for the documents in *test2000* and have calculated accuracy for each rank (light gray bars). The correct classification in both cases is almost always contained in the top four ranked classifications.

Some of the documents have multiple correct classifications. The articles with multiple correct classifications were removed, and accuracy was recalculated for each rank (dark gray bars). While the accuracy of the highest rank prediction is only slightly reduced from 72.8% to 72.0%, the accuracies of the second and third ranked classes is somewhat more reduced from 17.7% to 13.7% and 6.2% to 4.2%, respectively.

confidence score we separated the *test2000* predictions into ten groups by confidence score. For those predictions with the highest confidence scores (ranging from 0.9 to 1) the classifier's prediction was correct 93% of the time (see Figure 8.4). For predictions with lower confidence scores, the accuracy was proportionately less; the algorithm appears to estimate low confidence predictions conservatively.

When classifications are imperfect, it is important that they be associated with confidence scores. The maximum entropy classifier assigns probabilities to each possible prediction. After sorting predictions by probability score, we observe that a code's prediction accuracy matches its probability rank (Figure 8.3). Thus, the probability of the predicted code can be used as a measure of confidence on the prediction (see Figure 8.4).

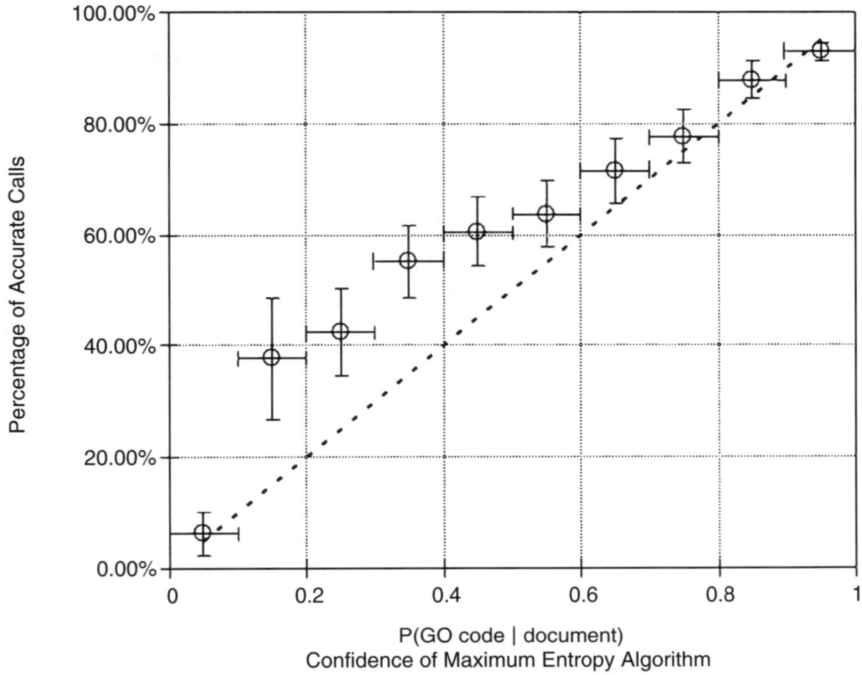

Figure 8.4 *Confidence scores are reliable indicators of accuracy.* Here we partitioned documents by the maximum entropy confidence score. The confidence score is the estimated probability of the predicted code given the document. For each partition we calculated accuracy on each subset. Each data point has an x-error bar indicating the size of the partition and a y-bar indication the 95% confidence interval on the accuracy estimate. As the confidence score increases along the x-axis the accuracy of the prediction increases proportionately. In fact the confidence score is a very reasonable predictor of the probability that the prediction is correct.

8.9 Annotating genes

At this point we have an accurate classifier. In addition, the classifications are associated with confidence scores. Given a set of documents relevant to a gene, the maximum entrorpy classifier can determine the relevant functional class of the documents. Then these classifications can be combined to determine the functional class of the gene.

Using the maximum entropy classifier we assign GO codes to each gene, based on their abstracts. We create a data set of abstracts associated with *S. cerevisiae* genes. Each gene is linked to a small set of abstracts; we use a set of curated abstracts for *S. cerevisiae* genes maintained by the Saccharomyces Genome Database at Stanford University. We make predictions only on those genes with three or more associated abstracts. There is a mean of 12.1 and a median of four articles per gene.

To validate our predictions we use the annotations assigned by the GO consortium. If an annotation was more specific than one in our set of 21, we map it back to a relevant ancestor based on the GO hierarchy. A total of 991 genes was annotated with GO codes relevant to this study by the consortium. In total, 835 genes were annotated and also had the requisite number of abstracts. We calculate the precision and recall at various thresholds for each of the annotations using the GO consortium assignments as a gold standard.

The voting scheme takes classifications of individual abstracts associated with a gene and combines them into a single gene classification. Maximum entropy classification provides the probabilities of a document's relevance to each of the 21 codes. The ad hoc parameter *fr* is the expected fraction of associated abstracts that should discuss a function if it is relevant to the gene. Here we selected a value of 1/3 for *fr* in all experiments; ideally a specific *fr* should be selected for each function separately depending on its prevalence in the literature. If *N* is the number of abstracts associated with the gene, analysis of each abstract with maximum entropy classification obtains *N* probability values for each GO code. We averaged the top ceil*(fr N)* probabilities for each code to score the code's relevance to the gene. This score ranged between 0 and 1; higher code scores indicated greater relevance to the gene. Genes with scores above a predetermined cutoff were assigned the code; the cutoffs were varied to create precision–recall plots. Other effective voting schemes are certinely feasible.

Here, we evaluate yeast gene annotation using abstracts. Even though many of the 21 categories that we study do not apply to yeast, we still include them in our calculations. Different thresholds of confidence can be used as a cutoff to assign an annotation to a gene. Typically, higher confidence values obtain higher precision at the cost of a lower recall. We computed the precision and recall for different confidence thresholds for each of the categories and plotted them in Figure 8.5. Ideally, precision remains 100% at all levels of recall.

Annotation efforts of genes from the curated set of abstracts yield uneven results. The precision–recall performance for some of the GO codes is reliable (Figure 8.5a), while others are passable (Figure 8.5b), and some are poor (Figure 8.5c). At one extreme, for the code "meiosis" we obtain the ideal precision–recall plot; a 100% precision was achieved at all levels of recall; in other words all of the correct genes were annotated. "Invasive growth" (16.7% precision at 100% recall), "sporulation" (100% precision at 11.1% recall) and "stress response" (9.1% precision at 76.9% recall) are the three codes that are difficult to annotate genes with.

Since the classifier performs consistently across all categories when the testing set and training set are similar (Table 8.6), the discrepant performance is explained by how well the training set represents the biological

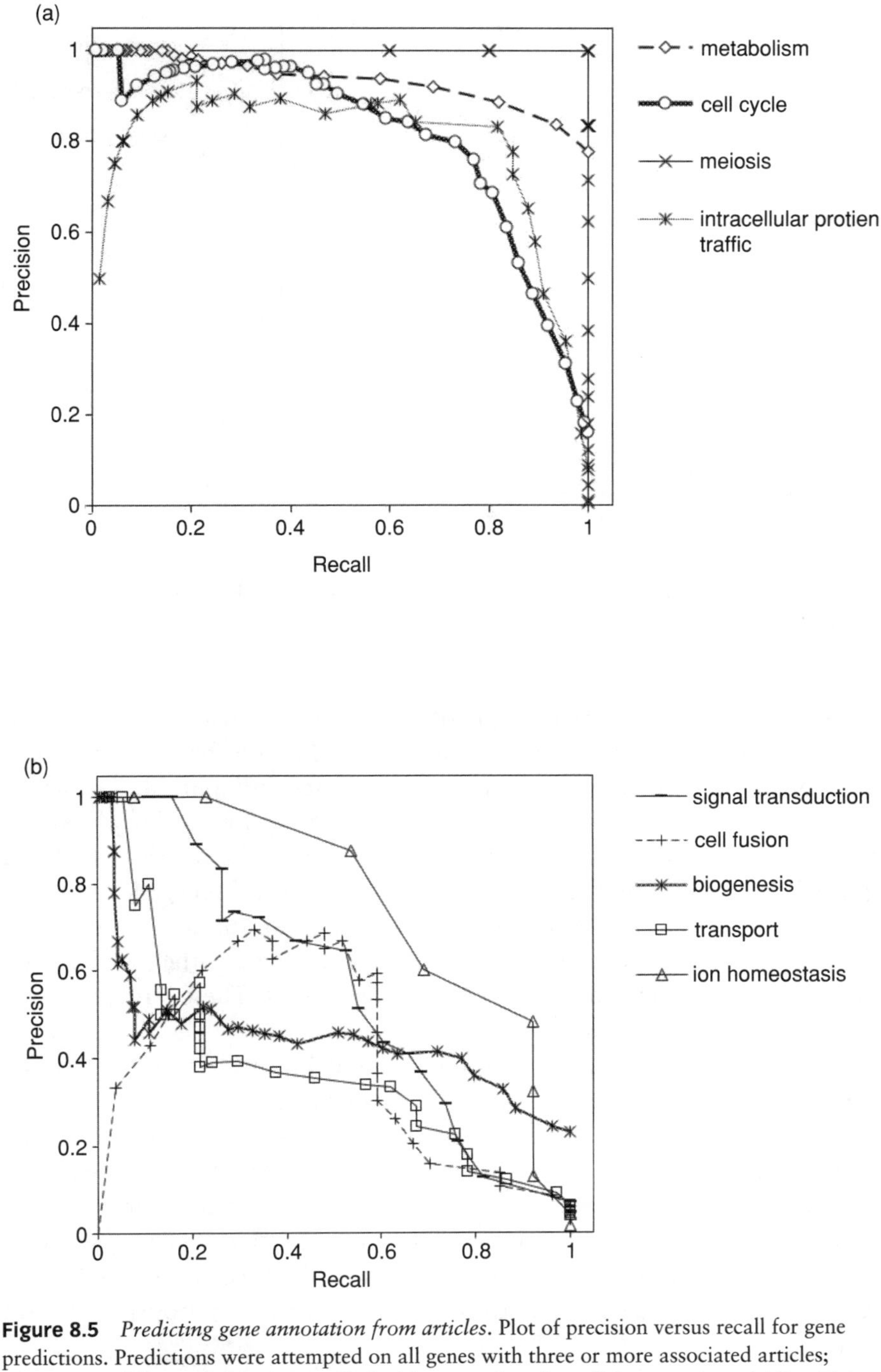

Figure 8.5 *Predicting gene annotation from articles.* Plot of precision versus recall for gene predictions. Predictions were attempted on all genes with three or more associated articles; correctness of the prediction was verified with annotations from GO consortium. (a) Precision–recall plot of the reliably predicted categories. (b) Precision–recall plot of the reasonably well predicted categories.

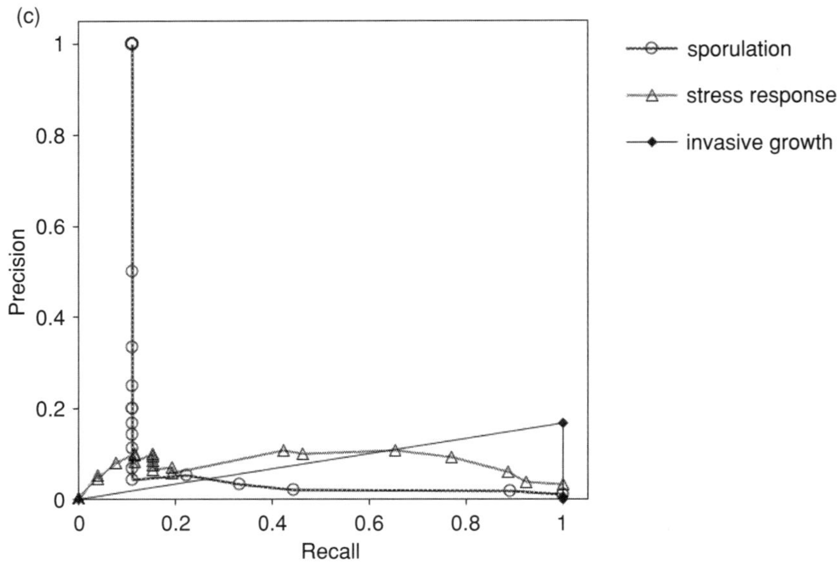

Figure 8.5 (*Continued*) (c) Precision–recall plot for categories for which predictions are poor. The predictions quality appears to correlate with the training set quality for that category.

processes. In general, the availability of a major MeSH heading corresponding to the code insures the quality of our PubMed search-based training set. Three of the four reliably predicted codes, plotted in Figure 8.5(a), had a single corresponding MeSH term, while two of the five codes plotted in Figure 8.5(b) had a single corresponding MeSH term. Of the three codes in Figure 8.5(c), one code, "invasive growth" had only a single gene, and thus a biased sample of text. For the other two codes, "sporulation" and "stress response", there are no corresponding MeSH terms for either, and ad hoc strategies were fabricated to create the set of articles. These strategies may be ineffective for these functions.

An ideal training set should be constructed by experts. The National Library of Medicine relies on experts to read the articles to assign MeSH headings (Bachrach and Charen 1978). These headings are likely to have a low false positive rate (high specificity) but may suffer from false negatives (low sensitivity) since experts assign some correct headings but may miss others. Our reliance on MeSH terms therefore assures that we get good training data when a MeSH heading corresponds directly to a GO code. However, the strategy is limited when there are no appropriate MeSH terms.

Better training sets consisting of more specific paragraphs from whole text articles and abstracts selected under expert supervision would address many of these difficulties.

References

Ashburner, M., C. A. Ball, et al. (2000). "Gene ontology: tool for the unification of biology. The Gene Ontology Consortium." *Nat. Genet.* **25**(1): 25–9.

Bachrach, C. A. and T. Charen (1978). "Selection of MEDLINE contents, the development of its thesaurus, and the indexing process." *Med. Inform. (Lond).* 3(3): 237–54.

Ball, C. A., K. Dolinski, et al. (2000). "Integrating functional genomic information into the Saccharomyces genome database." *Nucleic Acids Res.* **28**(1): 77–80.

Cherry, J. M., C. Adler, et al. (1998). "SGD: Saccharomyces Genome Database." *Nucleic Acids Res.* **26**(1): 73–9.

Funk, M. E. and C. A. Reid (1983). "Indexing consistency in MEDLINE." *Bull. Med. Libr. Assoc.* **71**(2): 176–83.

Hersh, W. R., R. T. Bhuporaju, et al. (2004). *TREC 2004 Genomics Track Overview.* The Thirteenth Text REtrieval Conference Proceedings (TREC 2004), Gaithersburg, MD, National Institute of Standards and Technology.

Kanehisa, M., S. Goto, et al. (2004). "The KEGG resource for deciphering the genome." *Nucleic Acids Res.* **32**(Database issue): D277–80.

Manning, C. M. and H. Schutze (1999). *Foundations of Statistical Natural Language Processing*. Cambridge, The MIT Press.

Mewes, H. W., D. Frishman, et al. (2000). "MIPS: a database for genomes and protein sequences." *Nucleic Acids Res.* **28**(1): 37–40.

Nigam, K., J. Lafferty, et al. (1999). "Using maximum entropy for text classification." *IJCAI-99 Workshop on Machine Learning for Information Filtering.*

Ratnaparkhi, A. (1997). A simple introduction to maximum entropy models. Philedelphia, Institute for Reseach in Cognitive Science, University of Pennsylvania.

Riley, M. (1993). "Functions of the gene products of Escherichia coli." *Microbiol. Rev.* 57(4): 862–952.

Rosenfeld, R. (2000). "Two decades of statistical language modeling: where do we go form here?" *Proceedimgs of the IEEE* **88**(8): 1270–1278.

Valencia, A., C. Blaschke, et al. (2004). Critical Assessment for Information Extraction in Biology (BioCreative), Granada, Spain. http://www.biomedcentral.com/1471-2105/6/S1/S1

9 Finding gene names

Successful use of text mining algorithms to facilitate genomics research hinges on the ability to recognize the names of genes in scientific text. In this chapter we address the critical issue of gene name recognition. Once gene names can be recognized in the scientific text, we can begin to understand what the text says about those genes. This is a much more challenging issue than one might appreciate at first glance. Gene names can be inconsistent and confusing; automated gene name recognition efforts have therfore turned out to be quite challenging to implement with high accuracy.

Gene name recognition algorithms have a wide range of useful applications. Until this chapter we have been avoiding this issue and have been using only gene-article indices. In practice these indices are manually assembled. Gene name recognition algorithms offer the possibility of automating and expediting the laborious task of building reference indices. Article indices can be built that associate articles to genes based on whether or not the article mentions the gene by name. In addition, gene name recognition is the first step in doing more detailed sentence-by-sentence text analysis. For example, in Chapter 10 we will talk about identifying relationships between genes from text. Frequently, this requires identifying sentences refering to two gene names, and understanding what sort of relationship the sentence is describing between these genes. Sophisticated natural language processing techniques to parse sentences and understand gene function cannot be done in a meaningful way without recognizing where the gene names are in the first place.

The major concepts of this chapter are presented in the frame box. We begin by describing the commonly used strategies that can be used alone or in concert to identify gene names. At the end of the chapter we introduce one successful name finding algorithm that combines many of the different strategies.

1) Using a gene name dictionary
2) Appearance of gene names
3) Syntax
4) Context of the gene name
5) Morphology of the gene name
6) Abbreviations for genes
7) Combined gene name finding method

9.1 Strategies to identify gene names

There are several commonly used approaches that can be exploited to recognize gene names in text (Chang, Shutze, et al. 2004). Often times these approaches can be combined into even more effective multifaceted algorithms. The first strategy is to recognize gene names by *dictionary*; this strategy involves using a predefined list of gene names to identify words in text as names. A second strategy is to use *appearance*; often gene names have common features or structure that can be exploited to determine whether a word is a gene name. A third strategy is to use the *syntax*; the fact that all gene names must be nouns or noun phrases can be used to eliminate many words as potential gene names. A fourth strategy is to look at the *context* of a word; that is, the words around a given word can help provide clues about whether or not it is a gene name. A fifth strategy is to look at the *morphology* of a word; different gene names often share a common root—recognizing those roots may help identify names themselves. Finally names often have *abbreviations* or are themselves abbreviations of a longer phrase; identification of these abbreviations can help provide clues about which words are true gene names.

If gene names were standardized and unique, the *dictionary* method would be most effective. Since this is rarely the case, more sophisticated name recognition methods using the other strategies are often necessary. Most of these methods use different combinations of these features to successfully identify gene names in text. We will discuss each of these features individually in this chapter.

9.2 Recognizing gene names with a dictionary

Perhaps the most reasonable way to start looking for gene names is to search for known gene names in the text. If gene names were standardized, unique, and used in a consistent manner in the literature, dictionary based look-ups would be all that was necessary for gene name recognition. The unfortunate challenge to this straightforward strategy is that gene nomenclature is quite complicated.

Most genome centers have a single standardized name for all genes in their organism. For example, the FlyBase name for the gene *breathless* is FBgn0005592. However, these names are used about as often as people refer to others with their social security numbers. They are neither intuitive nor descriptive and have no presence in common parlance. Often times the genome centers provide an official name, in this example *breathless* is the name of the gene. However, this official name is also a matter of consensus and opinion, there are frequently other common names that are used in the literature.

Most genes have several colloquial names coined by investigators. Often multiple names have arisen as a historical anomaly since the same gene might have been investigated or discovered by different groups under different contexts. Also abbreviations and variations of those names might be found commonly in the literature. In addition, there are frequently multiple standardized names from different versions of multiple controlled vocabularies.

In Table 9.1 we present the listed synonyms for two well-known fly genes, *breathless* and *heartless*. Some of the synonyms are based on standardized nomenclature while others are different names that have arisen for that gene throughout history. Some are related to their biochemical function, such as *fibroblast growth factor receptor*, while others are based on the phenotype of the gene when it is absent, such as *breathless*.

Extensive lists of synonyms have been compiled for many different organisms. FlyBase makes synonyms for fly genes available (Gelbart, Crosby et al. 1997). The synonyms in Table 9.1 are taken from the FlyBase synonym list. In Figure 9.1 we display a histogram of the number of synonyms per gene. It is a highly skewed distribution, where a few well-studied genes also have large numbers of synonyms, but the vast majority of the genes have very few names. In fact 44% of the recorded synonyms are pertinent to only 10% of genes. In Figure 9.2 we plot the relationship between number of articles and the number of synonyms; there is a striking correlation between these two variables. The log of the number of gene references and the total number of gene synonyms has a correlation of $r = 0.67$. Gene name recognition is therefore especially critical to mining the literature of well-studied genes; these genes have the greatest variety in gene names and also the most valuable available scientific literature. Similar synonym lists are available at many different genome database sites including Saccharomyces Genome Database, Mouse Genome Database, and Wormbase (Cherry, Adler et al. 1998; Stein, Sternberg et al. 2001; Blake, Richardson et al. 2002).

When compiled lists of synonyms are available, the task of the gene name finding can be as seemingly simple as string matching on text. However, there are some basic challenges to this approach. The synonym lists are manually curated, and are perpetually in a state of flux. It is difficult to keep such large lists current and gene names might be missing. Often gene names are used in text that are slightly varied from the listed synonym, and string matching can miss these occurrences. As an example, we might not be surprised to find *breathless* referred to in the scientific literature as FGFR-1; while this is similar to many of its synonyms, it is not an exact match. In addition gene names can be ambiguous since multiple genes can have the

Table 9.1 *Gene synonyms for two Drosophila genes.* Here we list the synonyms listed by the FlyBase Consortium for two closely related *Drosophila* genes: *breathless* and *heartless*. Some of the names correspond to the mutant phenotype of the gene; for example a mutation in *breathless* causes a deficiency in the fly respiratory system. Other names correspond to the biochemical function of the gene protein product; for example *breathless* is a fibroblast growth factor receptor. Other names are derived from standardized gene nomenclatures; for example CG6714 is a standardized name for the *breathless* gene.

Breathless	Heartless
0844/01	CG7223
breathless	CT22273
Btl	CT39172
BTL/FGFR2	DFGF-R1
CG32134	DFGF-R2
CG6714	DFR1
CT20816	Dfr1
dev	DFR-1
devenir	Dfr-1
D-FGFR	DFR1/DFGF-R2
DFGF-R1	DmHD-38
DFR2	DPR3
Dfr-2	dtk1
DmHD-311	Dtk1
dtk2	DTRK(FR1)
Dtk2	EMS2
FGF receptor	FGF receptor
FGFR	FGFR
fgf-r	FGFR2
Fgf-r	FGF-R2
Fgf-r: Fibroblast-growth-factor-receptor	Fibroblast growth factor receptor
FGFR1	Fibroblast growth factor receptor 1
Fibroblast-growth-factor-receptor	FR1
HD-311	Fr1: Fibroblast growth factor receptor 1
l(3)00208	FR1: Fragment 1
Tk2	HD-38
	heartless
	Htl
	HTL/FGFR1
	i100
	i150
	i79
	j372
	Tk1

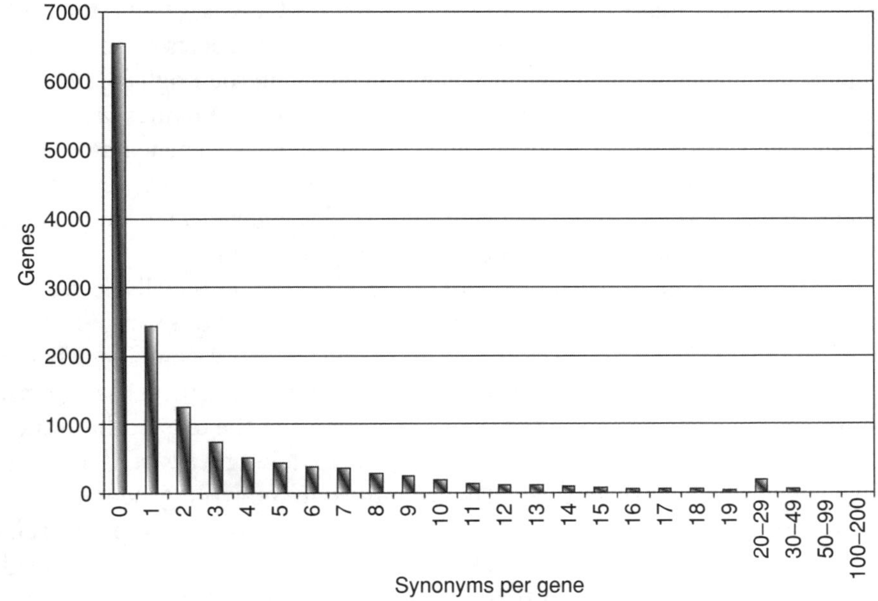

Figure 9.1 *Histogram of synonyms per gene.* This figure is a histogram of the number of synonyms per gene name. We looked only at a set of 14,385 fly genes known to have references. Of these, 45% of genes do not have a single name. The number of names per gene tapers off rapidly. About 44% of the gene names are assigned to 10% of the genes. About 90% of the gene names are assigned to 50% of the genes. These disparities are even more dramatic if we include the genes without any literature references.

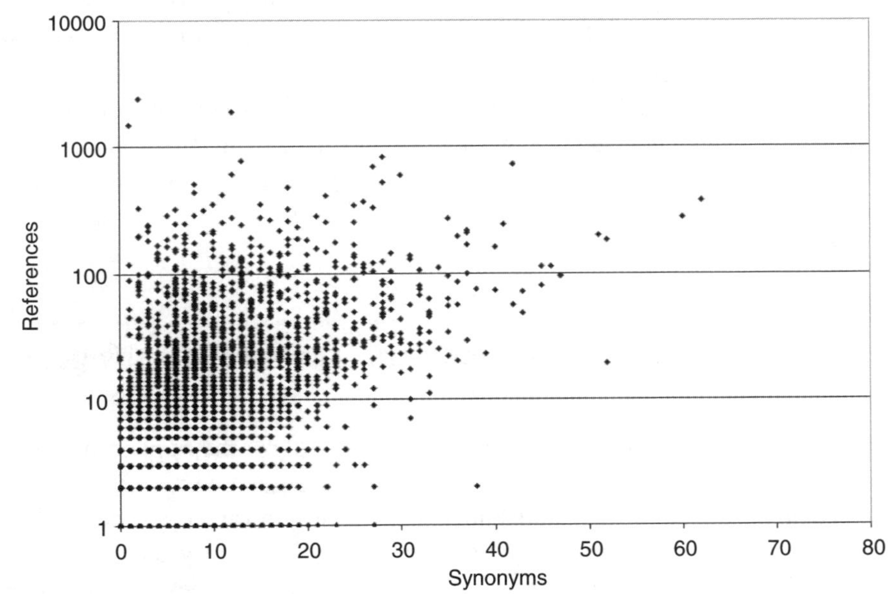

Figure 9.2 *Number of references versus number of synonyms.* Here each spot represents a gene. On the x-axis we plot the number of synonyms for that gene, and on the y-axis we plot the number of references for that gene on a log scale. There is a correlation of 0.67 between these two variables. The best-studied genes have the most synonyms.

same name. In Table 9.1 we can readily see that FGFR is a name for both *breathless* and *heartless*. One of the most challenging aspects of gene name recognition is that some names are common words in the English language that have non-biological meanings; for example, the gene names *breathless* and *heartless* can be used in other contexts besides that of a gene name. This challenge in gene name recognition is underscored by the fly gene *calcineurin*, which has *can* as a listed synonym, and the gene *early*. Simple text matching will in these cases turn up many false positives.

Nonetheless, this approach can be quite effective, especially if we are looking for names corresponding to a small set of genes. For example, given an article that is known to pertain to four specific genes, we can identify the exact location of references to individual genes in the text using synonyms. However, if we were scanning the same text for all of the tens of thousands of gene synonyms available using simple string matching we would obtain a large list of gene names in text, many of which are not true gene names.

One group evaluated how accurately string searches can match words in text to *drosophila* genes (Morgan, Hirschman et al. 2004). They concluded that matching with synonym lists combined with basic filtering strategies can be quite an effective approach to finding gene names in text. Matching for gene names on the full text without any filtering obtains only a 2.9% precision and 95% recall; there is an enormous number of false positives. Filtering out those gene name synonyms that (1) correspond to multiple gene names, (2) are not different from common English words, and (3) have greater than two characters in the name, improves the performance to 50% precision and 72% recall on the same data set. One other strategy to increase the sensitivity of dictionary-based gene name recognition algorithms might be to use approximate string matching instead of exact string matching. One group tested this approach; they used BLAST (see Chapter 2) to search scientific text to do approximate gene name matches to the scientific text (Krauthammer, Rzhetsky et al. 2000).

9.3 Using word structure and appearance to identify gene names

One approach to recognizing gene names is to analyze the appearance of a word or phrase to determine whether or not it is a gene name. Often there are specific clues that are a consequence of gene naming conventions that can serve as powerful hints in gene name recognition. In yeast genes, for example, there is a strong naming convention; these names are typically short three letter strings with a number at the end. Similarly the cytochrome P450 genes have a very standardized nomenclature that can be identified effectively with simple regular expressions. However, in general gene and protein names can become quite complicated. But other rules about word appearance can guide us in identifying gene names.

Chang and colleagues identified some of these clues in their investigation of gene name identification (Chang, Schutze et al. 2004). One of the most salient clues is the presence of the suffix "-ase". Examples of words frequently contained in gene names are "kinase", "phosphorylase", "transferase", and "topoisomerase". The suffix "-ase" in biology implies a protein product of a gene with catalytic properties. This group identified only 196 words in the English language that end with "-ase" that are non-gene names; all other words that end in "-ase" are, in fact, very likely gene names. Another less salient clue is the presence of the suffix "-in"; for example "insulin", "myosin", "actin" and "tubulin". Other useful hints include the use of lower and upper case letters, the presence of digits, the length of the word, and the presence of keywords found frequently in gene and protein names. Many gene names are shorter and they include numbers and capital letters. Some longer gene names may include keywords such as "activator", "receptor", and "transporter"; looking for these keywords in text could clue us in on gene names.

One of the pioneering studies utilizes word appearance to identify protein names (Fukuda, Tamura et al. 1998). This method uses a series of ad hoc rules to identify gene and protein names. The method is summarized in Table 9.2. In the first step the method looks for keywords and labels them as "feature terms". In the second step the authors identify words that have the appearance of atypical words that look like gene names; they refer to these terms as "core terms". They use aspects of the word such as length, presence of digits, and the presence of special characters. The authors use these terms to piece together gene names. The authors achieved a 98.8% recall and 94.7% recall with this strategy on a limited evaluation of their method. Another group took a similar approach to finding gene names with comparable results (Proux, Rechenmann et al. 1998).

9.4 Using syntax to eliminate gene name candidates

One simple, but key observation is that gene names are nouns or noun phrases. This offers a simple and easy way to remove many words as potential gene names. An off the shelf part-of-speech tagger can be used to identify the parts of speech for each word in a sentence. All words that are not in noun phrases can be immediately eliminated. In addition once a word that is likely part of a gene name is identified, part-of-speech tagging can be used to identify the entire the noun phrase containing it, and establish the complete gene name.

Table 9.2 *Summary of gene/protein name finding algorithm by Fukuda.*

Step 1. "Feature" terms are predefined

1. Include terms commonly found in protein names such as:
    a. Protein
    b. Receptor
    c. Domain
    d. Kinase
    e. Enzyme

Step 2. Identify core terms in text

1. Extract all words with capital letters, numerical figures, or special characters.
2. Eliminate those words >9 characters, all lower case, and with "-".
3. Eliminate words if >1/2 of its characters are special characters.
4. Exclude numerical words with units following (such as nM, fold, etc.)
5. Exclude all words that are part of the reference (authors name, journal, etc.)

Step 3. Concatenation

1. Concatenate any adjacent feature terms from Step 1 in the text, and core terms identified in the text in Step 2. This defines phrases that are either single terms, or multiple concatenated terms.
2. Apply the Brill part-of-speech tagger to identify the part of speech for the text.
3. Connect non-adjacent phrases if all words between them are nouns, adjectives, or numbers,
4. Extend all phrases to the left if there is a preceding determiner or preposition. The phrase should be extended just short of the determiner or preposition, and should be a noun phrase.
5. Extend all phrases to the right if they are followed by a Greek letter or single upper case letter.

Step 4. Resolve dependencies

1. Phrases can be concatenated despite comma, "and", and "of" with the use of several patterns.
2. If A, B, C, D, and E are phrases then concatenation can occur if the following forms are observed:
    a. A, B, ... C, and D feature term
    b. A, B, ... C, and D of E
    c. A of B, ... C and E
    d. A feature term, core term, and core term
    e. A of B
    f. A, B

Step 5. Filter

1. Remove phrases that are single feature terms.
2. Remove phrases for which the last word in not a noun.

We will not discuss the details of part-of-speech tagging except to note that it is a very mature subfield in natural language processing, and that high quality off the shelf part-of-speech taggers are available. Modern part-of-speech taggers routinely achieve greater than 95% accuracy in tagging (Brill 1994).

The method proposed by Fukuda uses, in addition to the appearance of the word, the part of speech. First a set of rules is used to identify likely gene name words based on appearance and other terms that frequently show up in gene names (see Table 9.2). Phrases are then merged together if the words between them are nouns, adjectives, or numbers. The authors are using parts of speech to construct noun phrases that represent the whole of the gene name. This key step requires the application of a part-of-speech tagger to determine the part of speech of the words between phrases.

9.5 Using context as a clue about gene names

Neighboring words around a putative gene name can help algorithms decide whether in fact it is a true gene name. In fact, the context of word use is a very valuable clue to human readers. For example consider the sentence:

The gene _____ is expressed under ...

Based on the presence and location of the word "gene" and "expressed" the reader would conclude that the blank space is an omitted gene name. Computational approaches can be devised that leverage the context as well.

For the purposes of computation, context might be defined as the words right before or right after a putative gene name, or a window of words before and after the putative gene name, or perhaps even all of the words in a given sentence.

The simplest strategy is to look at the word right before or right after a gene name and to see if matches certain pre-defined keywords. Chang and colleagues used chi-square test to identify a list of very effective keywords; we have listed these words in Table 9.3. Rindflesch defined a set of trigger words that frequently occur in noun phrases that contain gene names (Rindflesch, Tanabe et al. 2000). In his method he first identified noun phrases with a part-of-speech tagger, and then assumed phrases containing terms such as *activated*, *expression*, *gene*, and *mutated* mark them as phrases referring to genes.

An alternative, and somewhat more sophisticated approach might be to use statistical models to differentiate the context of a gene name from the context of non-gene names. One strategy might be to build a naive Bayes classifier to assess the word previous to and after a gene name. In this case, using Bayes' theorem and the independence assumption we assert:

$$P(w = G|\text{Context}) \sim P(\text{Context}|w = G) = P(\text{prev}|w = G)P(\text{next}|w = G)$$

Table 9.3 *Context trigger words for gene names.* These words were assembled by Chang et al. Using chi-square goodness of fit they assessed whether each word predicted a gene name before or after that word. Words that are asymmetrically present or lacking in the context of a gene name are listed below. In addition published p-values based on chi-square goodness of fit are listed below.

	Previous word	p-value	Next word	p-value
These words appear with gene names				
	gene	1.7 E-10	gene	0.0 E-00
			mrna	1.2 E-20
			protein	4.8 E-13
			promotor	1.3 E-13
			genes	1.5 E-10
			expression	4.5 E-09
			transcripts	3.8 E-08
			mrnas	3.4 E-07
These words do not appear with gene names				
	or	3.3 E-27	or	1.8 E-16
	by	3.5 E-21	were	9.0 E-09
	with	2.3 E-12	to	8.2 E-09
	to	1.5 E-11		
	in	1.5 E-10		
	for	2.0 E-08		

Here w is the word or phrase being assessed to see if it is a gene name, G. The higher this value, the more likely it is a gene name. With this equation we are asserting that the probability of the context of a gene name is the product of the independent probability of the word just previous to and just after the gene name. These probabilities can be calculated from training data. The probability distribution of words before and after gene names can be calculated separately. Sparse data may require the use of pseudo-counts to avoid zero probability words. This approach would be similar to the approach outlined in Chapter 8 for document classification, only in this case we are classifying context. For comparison we would need to define the probability of a context given that it was not a gene name:

$$P(w = \overline{G}|\text{Context}) \sim P(\text{Context}|w = \overline{G}) = P(\text{prev}|w = \overline{G})P(\text{next}|w = \overline{G})$$

where $\overline{G}$ represents a non-gene name. The ratio of these values predicts whether the context is more consistent with the context surrounding a gene name or a non-gene name.

Equivalently, maximum entropy or other text classification methods can be used to differentiate context surrounding a gene versus context surrounding a non-gene. This strategy can be further extended to include larger windows around the gene name instead of just the word just before and after the name.

Morgan and colleages explored an even more sophisticated method. They tested the use of hidden Markov models to examine the context of gene names (Morgan, Hirschman et al. 2004). Hidden Markov models (HMMs) are described in greater detail in Section 2.3 for the purpose of sequence analysis. In analyzing text, HMMs can be used in exactly the same way. The only difference is that instead of observations being amino acids, they are vocabulary words in the sentence. Hidden states in this case are the type of word observed. The hidden states or word types might include a state for "gene name" or for "gene context". The hidden Markov model can be trained on sentences containing gene names. Other potential gene names can be identified by scanning sentences with the hidden Markov models. Based on context some words may be assigned the hidden state "gene name". Morgan's method involved first tagging putative gene names with a hidden Markov model, and then confirming the gene name with a dictionary look-up. The authors found that the HMM could be used to filter dictionary look-ups to achieve 88% precision at 61% recall if they removed ambiguous gene names. This represents a substantial improvement over dictionary look-ups with the filtering scheme investigated in the same study described in Section 9.2.

9.6 Morphology

An interesting facet about gene names is that they are often derived from a common root. This is particularly the case when genes are part of a larger family of genes. Many genes will have the same root, but different prefixes and suffixes. Recognizing these roots in the text can be a critical clue in recognizing gene names. For example consider the root *ank* for example. In mouse there are three genes with the morphological root *ank* : *ank1* (ankyrin 1, erythroid), *ank2* (ankyrin 2, brain), and *ank3* (ankyrin 3, epithelial). Recognizing the root *ank* would help identify any of these gene names in text.

To date, word morphology has not been extensively exploited in gene name recognition.

9.7 Identifying gene names and their abbreviations

In biology, professionals use gene name abbreviations liberally, so much so that the abbreviation can often become the common name of the gene. For example, the *breathless* gene is also known as "fibroblast growth factor receptor"; this longer name describes one of the important functions of this gene. Commonly, in the literature it is referred to as FGFR, which is its abbreviation. Another example is “cyclin dependent kinase” or CDK. Previously we mentioned the gene “calcineurin” which is abbreviated as “can”. Biology is filled with countless examples of many different types of abbreviations. In cases such as this, recognizing abbreviations in the text as well as their longer form can be a critical component to gene name recognition.

Several groups have proposed algorithms to identify abbreviations in the literature (Yoshida, Fukuda et al. 2000; Pustejovsky, Castano et al. 2001; Chang, Schutze et al. 2002; Yu, Hripcsak et al. 2002). These algorithms can be used to mine large collections of literature, such as PubMed to create lists of common abbreviations. Here we will focus on a very effective strategy proposed by Chang and colleagues that uses dynamic programming to link phrases and their abbreviations (Chang, Schutze et al. 2002).

The method has four steps: (1) scanning the text for phrases and their putative abbreviations; (2) aligning the candidate abbreviation to the abbreviated phrase; (3) creating a feature vector that quantifies different aspects of the putative abbreviation; and (4) using machine learning to score the putative abbreviation with the feature vector.

In the first step, we begin by scanning the text for parentheses. The text within the parentheses is regarded as a putative abbreviation for the phrase preceding the left parenthesis (see Figure 9.3). We remove candidate abbreviations that are more than two words long and those candidates without any letters. To limit the scope of the search, we filter potential abbreviated phrases only up to the previous comma or semicolon. To identify the phrase that the candidate abbreviation might be abbreviating, we take at most the

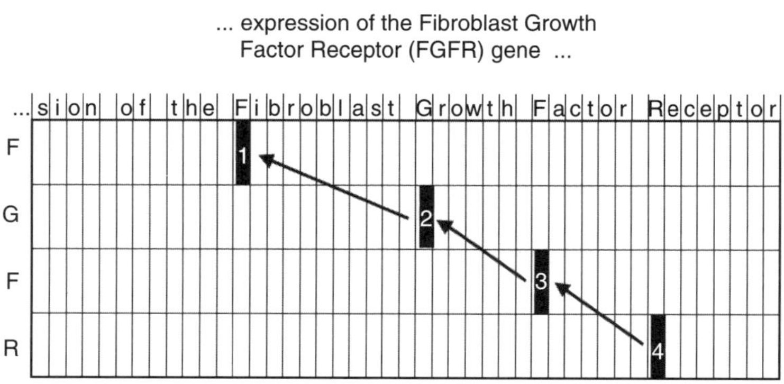

Figure 9.3 *Finding abbreviations.* The method of abbreviation discovery proposed by Chang and colleagues involves first finding putative abbreviations in parentheses. The words prior to the left parenthesis are presumed to be the phrase that is abbreviated. The abbreviation and prior words are aligned with dynamic programming. Unlike sequence analysis, there are no gap penalties or approximate matches. Note there is some potential ambiguity in matches: the R in “FGFR” could have matched the last R in “Receptor” or the last R in “Factor”.

$3N$ words preceding the left parenthesis, where N is the number of letters in the candidate abbreviation.

In the second step we use dynamic programming to align the putative abbreviation with the abbreviated phrase. Dynamic programming for string alignment is described in the context of DNA and protein alignments in Section 2.3. The approach is identical here. The only difference is that there is no gap penalty for introducing spaces. The score of a match between two letters is 1, and the score of a mismatch between two letters is 0. An example of this sort of alignment is provided in Figure 9.3. As with dynamic programming for sequence alignment, once the scoring matrices are defined, trace-back through the matrix from the highest scoring position can be used to determine the optimal alignment between the phrase and the abbreviation.

Given an alignment between an abbreviation and abbreviated gene name, we evaluate the alignment against eight features. These features are described in detail in Table 9.4. Each alignment is automatically assessed and scored for each of these features.

Chang and colleagues trained a binary logistic regression classifier to score these alignments based on the feature values. Logistic regression is another supervised machine learning binary classification method similar to linear discriminant analysis described in Section 2.4. If the probability of an alignment being a true abbreviation and abbreviated phrase is p, logistic regression assumes that the log odds ratio is a linear function of the feature set.

Table 9.4 *Features to evaluate abbreviations.* These are the features proposed by Chang and colleagues to assess abbreviations and abbreviated phrase alignments. Once an alignment is scored with the system below, logistic regression is used to determine whether the alignment represents a true abbreviation. The logistic regression parameters are included in the right column.

Feature	Description	β
0	Constant (β_0)	−9.70
	Abbreviation features	
1	Percent of lower case letters	−1.21
2	Percent of letters in the abbreviation that are aligned	3.67
	Abbreviated phrase	
3	Percent of letters aligned at the beginning of a word	5.54
4	Percent of letters aligned at the end of a word	−1.40
5	Percent of letters aligned on a syllable boundary	2.08
6	Percent of letters aligned immediately after another letter	1.50
7	Number of words not aligned to the abbreviation	−5.82
8	Average number of aligned letters per word	0.70

$$\log\left(\frac{p}{1-p}\right) = \beta_0 + \sum_i \beta_i x_i$$

The left-hand side is the log odds ratio; the β_0 parameter is a normalization constant; the β_i parameters are the feature weights in Table 9.4; and x_i are the actual values for the individual features. The investigators trained the logistic regression equation on a training set of 1000 examples to obtain the parameters listed in Table 9.4. The investigators obtained 95% precision at 75% recall.

This method has been used to scan the MedLine database to assemble all abbreviations. In gene finding tasks, these abbreviations can be used to identify abbreviated gene names in the text. To determine whether a gene name is being described one could either consider the actual word in the text, or see if it corresponds to an abbreviation or expanded form that looks more like a true gene name.

9.8 A single unified gene name finding algorithm

As we have already suggested, many of these methods use multiple strategies to recognize gene names. For example, Fukuda and colleagues rely heavily on the appearance and part of speech (see Table 9.2). We also referred to Morgan and colleagues; their method relied heavily on a dictionary, but used context to optimize their results. The above approaches cannot be used effectively in isolation, and require some degree of combination to effectively find gene names.

Optimal combinations will depend on the exact application, and may even vary from species to species. For example, while *dictionary* might be a powerful approach in yeast where names are reasonably standardized, it might be much less effective in humans where the names are much less standardized.

In this section we will outline the method proposed by Chang and colleagues (Chang, Schutze et al. 2004). This algorithm represents only one of countless gene name finding algorithms that have been proposed at this point. Unfortunately, it is not practical for us to delve into the details of all proposed gene finding algorithms. Our goal here is give an example of the way in which these different features can be combined into a gene name identification algorithm.

This method illustrates usage of all of these strategies except *dictionary*. This method proceeds through five steps: (1) tokenize, (2) filter, (3) score, (4) extend, and (5) match abbreviation. The final product is a score that predicts the likelihood that the phrase is a true gene name.

In the first step, text is segmented into sentences, and further segmented into words. To identify words, the algorithm separates the text in sentences by tokenizing on spaces and punctuation in the text, except for dashes. Dashes are word boundaries unless the previous token is a single letter, or the next token is a roman numeral or number.

The second step, filter, uses syntax. A part-of-speech tagger is applied to the sentences to assign a part of speech to each of the words. Only the words that are identified as nouns, adjectives, participles, proper nouns, and foreign words are kept for further consideration. In addition numbers, roman numerals, virus names, common chemical compounds, organism names, and a list of common technical words are also removed.

In the third step, the remaining words are scored. Prior to scoring, two special cases are recognized and treated separately. Words that end in "-ase" that are not part of a predefined collection of 196 non-gene name words receive the highest possible score. Similarly words that conform to the standardized nomenclature of cytochrome enzymes also receive the highest possible score. The remaining words are examined for features that describe their morphology, appearance, and context.

To score the appearance of a word, key binary features are formulated; these features are described in Table 9.5. These particular features were selected because of their presumed ability in distinguishing gene names from non-gene names. If a candidate word has a feature, the number 1 is assigned to that feature in a feature vector; if it lacks that feature, 0 is assigned to the feature vector. In addition to these binary features a final appearance feature was defined that pertained to words containing the "-in" suffix. The authors hypothesized that those words with the suffix that are gene or protein names have a recognizable pattern of letters. The authors utilized an *n*-gram classifier that was trained on known gene names that had the "-in" suffix and also on non-gene names (Smarr and Manning 2002). The *n*-gram classifier is a word statistical model that for any word with the "-in" suffix will assign a value proportional to the probability that the word is indeed a gene name. Given a candidate word with the "-in" suffix, the *n*-gram classifier is used to obtain a predictive value. That value is entered as a final appearance feature. Words that lack the "-in" suffix are assigned 0 to this feature.

The method also defines features that score the morphological root of the word. Eight features are defined, each representing a different type of variation (see Table 9.6). Then for each of the eight different kinds of variation we calculate the number of that kind of variation:

$$\max\left[\log\left(\frac{1}{1000}, \frac{\#\text{Vars}}{\#\text{Root}}\right)\right]$$

So if *cdc* occurs alone in a corpus of 30 times, and *cdc+number* (such as *cdc8* or *cdc22*) occurs 300 times, then the value for the morphological feature score for *cdc* is max[−3,1]. The 1/1000 ratio is included in the

Table 9.5 *Features to characterize the appearance of a word.* These features are defined for each of the candidate gene name words. All of these features are binary features.

Length
1 letter
2 letters
3–5 letters
>6 letters
Presence of numbers
First letter is digit
Last letter is digit
Last letter is a Roman numeral
Cases
Entire word is capitalized
Last letter is upper case
Mixed upper and lower case word
Word ends with upper case letter and number
Other
Presence of Greek letter
Presence of dash

Table 9.6 *Different morphological variants for a root.* Each of these types of morphological variants is scored as described in the text. The score for each of the variants is used as a feature vector to score candidate gene names.

root + Greek letter
Greek letter + root
root + Roman number
"apo" or "holo" + stem
stem + upper case letter
stem + number
stem + upper case letter + number
stem + lower case letter

equation to normalize for aberrant misspellings in the corpus. Given a candidate word, the root is identified, and feature values are assigned with the above equation. If there is ambiguity about what the root in the word is, the highest scoring root is utilized.

To calculate the context of a word, look at the word just before and just after the candidate word. For each trigger word listed in Table 9.3 we define a feature. The total number of times each trigger word appears with the candidate word across the corpus is entered as the value for that feature.

So in total we have defined a feature vector with 14 appearance features, eight morphological features, and 18 context features. For each candidate word this defines a total of 40 features. The authors tested naive Bayes, maximum entropy, and support vector machines as machine learning methods to classify candidate gene names based on these feature vectors. They trained each of these different types of classifiers on known gene names, and known non-gene names. The score that these classifiers assigned to a given word was used to predict whether the potential gene name that they were a part of was in fact a true gene name. Higher scores were considered more predictive of true gene names.

The authors then extend the gene name from the candidate word to include nouns, adjectives, and participles preceding the candidate word and Roman numerals, Greek letters, and single letters proceeding the candidate word.

In the final step, the algorithm checks if the gene name is an abbreviation for a longer gene name, or if it has an abbreviation using the method to identify abbreviations described in the previous section. If the abbreviation has the higher score, that score is transferred to the gene name. Alternatively if the gene name is an abbreviation, and its long form has a higher score, that score is transferred.

The authors found that maximum entropy classification and support vector machines offer comparable performance. The optimized support vector machine classifier obtains 83.3% recall at 81.5% precision. Most importantly, however, they showed that removal of any of the different strategies used in the program markedly decreased performance. That is to say that all of the incorporated strategies were actually critical to the algorithm's high performance level. At a fixed recall of 75%, removing different modules causes a loss in precision ranging from 2% loss (by removing context features) to 23% loss (by removing part-of-speech filtering). Examination of all of these features together contributes to the overall performance of the algorithm.

We anticipate that the future will bring novel combinations of these different aspects to continue improving gene name identification. This certainly remains a dynamic and very active area of research.

References

Blake, J. A., J. E. Richardson, et al. (2002). "The Mouse Genome Database (MGD): the model organism database for the laboratory mouse." *Nucleic Acids Res.* **30**(1): 113–5.

Brill, E. (1994). Some advance in transformation based part of speech tagging. *Proceedings of the 12th National Conference on Artificial Intelligence (AAAI-94)*, pp. 722–7.

Chang, J. T., H. Schutze, et al. (2002). "Creating an online dictionary of abbreviations from MEDLINE." *J. Am. Med. Inform. Assoc.* **9**(6): 612–20.

Chang, J. T., H. Schutze, et al. (2004). "GAPSCORE: finding gene and protein names one word at a time." *Bioinformatics*. **20**(2): 216–25.

Cherry, J. M., C. Adler, et al. (1998). "SGD: Saccharomyces Genome Database." *Nucleic Acids Res.* **26**(1): 73–9.

Fukuda, K., A. Tamura, et al. (1998). "Toward information extraction: identifying protein names from biological papers." *Pac. Symp. Biocomput*: 707–18.

Gelbart, W. M., M. Crosby, et al. (1997). "FlyBase: a Drosophila database. The FlyBase consortium." *Nucleic Acids Res.* **25**(1): 63–6.

Krauthammer, M., A. Rzhetsky, et al. (2000). "Using BLAST for identifying gene and protein names in journal articles." *Gene.* **259**(1–2): 245–52.

Morgan, A. A., L. Hirschman, et al. (2004). "Gene name identification and normalization using a model organism database." *J. Biomed. Inform.* **37**(6): 396–410.

Proux, D., F. Rechenmann, et al. (1998). "Detecting gene symbols and names in biological texts: a first step toward pertinent information extraction." *Genome Inform Ser Workshop Genome Inform* **9**: 72–80.

Pustejovsky, J., J. Castano, et al. (2001). "Automatic extraction of acronym-meaning pairs from MEDLINE databases." *Medinfo.* **10**(Pt 1): 371–5.

Rindflesch, T. C., L. Tanabe, et al. (2000). "EDGAR: extraction of drugs, genes and relations from the biomedical literature." *Pac. Symp. Biocomput*: 517–28.

Smarr, J. and C. Manning (2002). "Classifying unknown proper noun phrases without context." *Technical Report, Stanford University.*

Stein, L., P. Sternberg, et al. (2001). "WormBase: network access to the genome and biology of Caenorhabditis elegans." *Nucleic Acids Res.* **29**(1): 82–6.

Yoshida, M., K. Fukuda, et al. (2000). "PNAD-CSS: a workbench for constructing a protein name abbreviation dictionary." *Bioinformatics.* **16**(2): 169–75.

Yu, H., G. Hripcsak, et al. (2002). "Mapping abbreviations to full forms in biomedical articles." *J. Am. Med. Inform. Assoc.* **9**(3): 262–72.

10 Protein interaction networks

Genes and proteins interact with each other in many complicated ways. For example, proteins can interact directly with each other to form complexes or to modify each other so that their function is altered. Gene expression can be repressed or induced by transcription factor proteins. In addition there are countless other types of interactions. They constitute the key physiological steps in regulating or initiating biological responses. For example the binding of transcription factors to DNA triggers the assembly of the RNA assembly machinery that transcribes the mRNA that then is used as the template for protein production. Interactions such as these have been carefully elucidated and have been described in great detail in the scientific literature.

Modern assays such as yeast-2-hybrid screens offer rapid means to ascertain many of the potential protein–protein interactions in an organism in a large-scale approach. In addition, other experimental modalities such as gene-expression array assays offer indirect clues about possible genetic interactions.

One area that has been greatly explored in the bioinformatics literature is the possibility of learning genetic or protein networks, both from the scientific literature and from large-scale experimental data. Indeed, as we get to know more and more genes, it will become increasingly important to appreciate their interactions with each other. An understanding of the interactions between genes and proteins in a network allows for a meaningful global view of the organism and its physiology and is necessary to better understand biology.

In this chapter we will explore methods to either (1) mine the scientific literature to identify documented genetic interactions and build networks of genes or (2) to confirm protein interactions that have been proposed experimentally. Our focus here is on direct physical protein–protein interactions, though the techniques described could be extended to any type of biological interaction between genes or proteins.

There are multiple steps that must be addressed in identifying genetic interaction information contained within the text. After compiling the necessary documents and text, the first step is to identify gene and protein names in the text. This can be difficult in itself; this issue is addressed in Chapter 9. The second step is to identify candidate sections of text, for example abstracts or sentences, where the names of both genes in the putative interaction co-occur. Often the co-occurrence of two gene or protein names alone is suggestive of some sort of relationship. The final step is to analyze that text and to determine whether an interaction is described, and what that interaction is.

The key concepts in this chapter are described in the frame box. We begin with a discussion of genetic and protein interaction networks. We proceed to describe the high throughput experimental modalities that have been recently popularized to derive protein–protein interactions. We then discuss the predictive value of gene name co-occurrence alone in abstracts or sentences. We then present some more advanced methods to analyze text segments for interactions; we describe and give examples of information extraction and statistical machine learning approaches to this problem.

10.1 Genetic networks

Genetic networks are graphical depictions of the relationships between different genes in the same organism. The simplest networks simply connect genes together through binary links with each other. These links may indicate an intimate physical connection between the genes or proteins implying a specific type of relationship. Alternatively links may have a more generic meaning; they may imply only that these two genes have some sort of vague functional connection to each other. More complicated networks encode many different kinds of specific interactions between genes. These binary interactions may have a directional component to them as well; proteins may interact with each other in a non-symmetric manner. In most cases we expect that genes that are functionally similar to be either directly connected or closely associated with each other in the network.

1) Genetic networks
2) Experimental modalities
   a) Yeast-2-hybrid method
   b) Affinity chromatography
3) Predicting protein–protein interactions from text
   a) Gene name co-occurrence
      i) Sentences
      ii) Abstracts
      iii) Number of co-occurrences
   b) Information extraction
   c) Statistical machine learning

Genetic networks can be defined to encode all kinds of different types of interactions between genes and proteins. To start with, consider protein–protein interactions. Proteins can interact with each other in many sophisticated and complicated ways. Experimental modalities such as yeast-2-hybrid assays can detect binding between proteins. The most generic protein–protein interaction networks simply have links between all genes whose protein products bind at any point and time to each other. However proteins can interact with each other in as many different ways as there are proteins. They may be bound together in a permanent structure as part of a complex, the way two ribosomal proteins might be tightly bound to each other. Many proteins are part of larger subcellular complexes that function collectively as units; these proteins are often bound to each other statically. Alternatively, proteins can bind to each other dynamically. They may come together briefly to achieve some functional purpose. For example, the transcription factor proteins bind DNA and each other when a gene is about to be transcribed. A large functional complex is formed that engages in DNA transcription. Proteins can also interact in an even more transient fashion. For example, they can chemically modify other proteins. For example, protein kinases attach phosphate groups to proteins and in the process can often modify the function of the protein. Protein phosphatases remove phosphate groups, on the other hand, and return the phosphorylated protein to its native structure and function. There are many functional modifications of proteins that are catalyzed by proteins.

Of course proteins can interact with genes. Transcription factors and promoters are proteins that influence the expression state of a gene. These moieties bind the DNA and can cause a gene to be expressed and ultimately translated into a protein. Studies with gene expression arrays in which a single transcription factor is artificially over-expressed can help determine which genes are affected by a particular transcription factor.

Many of the known interactions have been tediously obtained by careful experimental biology, and are documented and described in the scientific text. Appropriate use of text mining strategies can be helpful in identifying and characterizing interactions.

Our emphasis in this chapter is protein–protein interactions specifically.

10.2 Experimental assays to identify protein networks

10.2.1 Yeast two hybrid.

One of the most widely used experimental methods to rapidly determine protein–protein interactions is the yeast-2-hybrid method (Zhu, Bilgin et al. 2003). This method can be scaled to do large screens. For example, the assay can be executed in an array format that facilitates rapid screening (Uetz 2002). Large yeast-2-hybrid screens to determine protein interaction maps have been undertaken in helicobacter pylori, yeast, worm, and fly (Uetz, Giot et al. 2000; Walhout, Sordella et al. 2000; Ito, Chiba et al. 2001; Rain, Selig et al. 2001; Giot, Bader et al. 2003).

This method relies on a reliable reporter gene; the reporter gene is a gene whose expression can be easily noted phenotypically. For example, one ideal reporter gene is beta-GAL; it expresses a fluorescent protein whose excessive expression can be noted macroscopically by a color change in the cell. To assess whether two proteins interact, one protein is fused to a DNA binding domain that binds upstream of a reporter gene in yeast. The protein fused to the DNA binding domain is also known as the "bait". The second protein is fused to a transcription-activating domain; this protein is known as the "prey". The theory is that if the bait and prey proteins interact, then the transcription-activating domain is brought close to the upstream region of the reporter gene when the bait and prey bind. The result is that the reporter gene is over-expressed. If the expected color change is observed in the cell, it is then assumed that the bait and prey proteins are interacting. While this method is fast and efficient, it has known limitations as well. Binding is assessed in the nucleus, and membrane bound proteins are unable to be assessed properly. Two hybrid screens are known to miss many accepted protein interactions; they also have a reasonably high false positive rate.

10.2.2 Affinity precipitation.

Another popular method to rapidly assess protein–protein interactions is affinity precipitation (Zhu, Bilgin et al. 2003). Two comprehensive studies in yeast have been successfully implemented (Gavin, Bosche et al. 2002; Ho, Gruhler et al. 2002). The general approach is to use molecular biological methods to attach bait proteins with a polypeptide tag. This tag is capable of binding a chromatography column. Then, cellular extracts are purified using the chromatography column. The tag and the attached protein bind the column. Proteins that bind the protein remain in the column, while the other proteins and cellular debris wash off. Harsher conditions are then employed to elute the binding proteins. Electrophoresis can be employed to sort proteins by mass into bands on a gel. Mass spectroscopy is one approach that can then be employed to determine the identities of the different proteins that were bound to the bait proteins.

10.3 Predicting interactions versus verifying interactions with scientific text

We address two separate roles of scientific text in the context of gene networks in this chapter. One possibility is using the text to learn networks of genes only from pre-established knowledge. This is tantamount to documenting all gene interactions described in the scientific text. Alternatively, we can use the corpus of published scientific text to verify experimentally predicted interactions. The simplest and most sensitive approach to both of these challenges is to count gene-gene co-occurrences and assume that if two genes co-occur that there is some sort of interaction between them.

These different uses of the literature imply different goals, however. Verification of predicted interactions typically requires methods that are very sensitive, at the cost of specificity. That is to say, given a prediction of a protein interaction, we are looking for any supportive evidence to say that the interaction might be valid. So co-occurrence in text may be the most appropriate strategy – as it is as sensitive as any text-based method can be expected to be. On the other hand, if the goal is to simply mine the text without any experimental information, then we would desire a more specific method. The goal of such a method would be to accurately obtain from the text well-documented interactions; we would want only the interactions that we could be certain of. So simple co-occurrence in text may lack the specificity requisite to build high quality networks – more sophisticated text mining strategies are required to determine which sentences are actually describing the interactions between the co-occurring proteins.

10.4 Networks of co-occurring genes

In either mining gene networks from the scientific text or assessing the validity of predicted interactions, looking at gene–gene co-occurrence in the literature is the initial step. For example if we are trying to assess whether or not the protein product of two genes physically interact, ideally we are looking for sentences such as:

Protein A binds and modifies the function of protein B.

In practice, however we are not often so fortunate. Sentences are written with subtle grammer and vocabulary, and understanding the content of such sentences computationally can be difficult. The first step is simply identifying sentences or paragraphs that contain the names of both genes. These sentences where gene names co-occur are good candidate sentences. The more sentences we find with the same pair of gene names, the more likely that one of them describes a physical interaction between them. In any case frequent co-occurrence might at the very least imply a vague functional relationship if not a specific interaction.

Jenssen and colleagues explored networks based on gene name co-occurrence in the scientific text (Jenssen, Laegreid et al. 2001). They looked for the names of 13,712 human genes in PubMed abstracts. They defined a link between two genes if they were mentioned together in the same abstract. They noted the number of abstracts that those two genes occurred together in as a measure of the strength of the interaction. They reported that 62% of the genes they examined had at least a single reference. Of those genes with references, 88% have at least one gene that it co-occurs with in the text. The average number of abstracts that each name pair was represented in was 7.8. To get a sense of the quality of the predicted gene links the authors sampled 500 gene pairs that co-occurred in five or more abstracts. Of these, 72% corresponded to true biological links between the two genes. Alternatively, doing the same experiment on gene pairs with a strength of one abstract reveals that only 60% correspond to true biological links. Incorrect links were always a function of failure to either recognize gene names correctly or associate them with the right gene. This study underscores the importance of high quality gene name recognition in text analysis. The authors provide the network as a conceptual map between genes, and do not provide an explanation for the links.

10.5 Protein interactions and gene name co-occurrence in text

Here we will assess how effective co-occurrence in the scientific literature is at predicting and verifying gene–gene interactions. Specifically, we look at a comprehensive collection of protein interactions, and see how well co-occurrence of gene names in text correspond to these interactions.

We use as a gold standard the General Repository for Interaction Datasets (GRID) (Breitkreutz, Stark et al. 2003). We focus only on affinity precipitation and yeast-2-hybrid predicted interactions as a gold standard. Comprehensive information about this data set is summarized in Table 10.1. There are a total of 4,621 yeast genes with 12,509 documented interactions. This means that of all possible gene pairs, about 0.12% correspond to documented physical interactions. This is by no means a perfect gold standard. These assays have high false positive and false negative rates, and many consider them questionably reliable. In fact the literature clearly describes many interactions between proteins that are not represented by GRID. We use it for lack of a better comprehensive experimentally derived standard. We must recognize that the sensitivities and

Table 10.1 *The General Repository for Interaction Datasets (GRID).* This table describes the data contained in GRID. Each row represents a specific sort of assay represented in GRID. For each assay we list the total number of genes examined, the total number of interactions found between those genes, and the total number of studies included in GRID.

GRID data set	Genes	Interactions	Studies
Total	4711	13,607	703
Affinity chromatography	172	134	98
Affinity precipitation	2365	6894	277
Biochemical assay	6	3	3
Dosage lethality	4	2	2
Purified complex	158	128	93
Reconstituted complex	5	5	4
Synthetic lethality	755	985	318
Synthetic rescue	6	5	2
Two hybrid	3873	6127	267

specificities that we derive may be gross underestimates. However, they do provide a nice starting point that helps to get a sense of the effectiveness of text-based strategies and compare different approaches.

Here we will assess how sensitive and specific simple co-occurrence is at identifying these interactions. Due to the limited availability and difficulty in obtaining full text, we looked for co-occurrence in article abstracts. This is a limited approach, as much of the valuable information about genes is summarized in the introduction and discussion of published papers. The yeast literature has the advantage of having many well-studied genes with abundant references.

We obtain the references from the Saccharomyces Genome Database. For each abstract we identify gene names in the text. We use a synonym list of gene names to identify gene names in each sentence. We look only for the names of genes that are referenced by that document according to the Saccharomyces Genome Database reference index. This is the dictionary strategy outlined in Chapter 9. First, we identify gene name co-occurrence in abstracts. There were some 287,533 co-occurrences among these abstracts. Of these, 133,687 co-occurrences are pertinent to the 4621 genes that we are focused on. We then also look at sentence co-occurrences. We broke up the abstracts into individual sentences. Then we select all sentences with two or more gene names. There were 24,524 sentences with co-occurring gene names from the set of 24,378 abstracts. Of these, 17,712 are co-occurrences where both genes are from the list of genes that were in the interaction data set. The results are summarized in Table 10.2.

Abstract co-occurrence can be a powerful indicator of a relationship between genes. Since there are many genes associated with each abstract there are many more co-occurrences than when looking at co-occurrences

Table 10.2 *Data about sentence and abstract co-occurrences.* The first row represents the total number of co-occurrences among the abstracts or sentences relevant to the proteins in the GRID data set. The next row lists the total number of gene pairs that the textual co-occurrences correspond to. The next row lists the percentage of those co-occurrences that correspond to protein interactions reported in GRID. The next set of rows lists the total number of interactions in GRID, the number and percentage that co-occur, and the average number of co-occurrences per interacting gene pair. The same series of data is provided for non-interacting gene pairs. Then we list the percentage of the co-occurring gene pairs that are actually interacting gene pairs. Finally we list the parameter R, the probability that an interacting gene pair co-occurs divided by the probability that a non-interacting gene co-occurs.

		Abstract	Sentences
Text	co-occurrences	133,687	17,712
	Pairs of co-occurring genes	74,504	5,708
	%co-occurrences that correspond to interacting genes	11.26%	26.15%
Interacting gene pairs	Total	12,509	12,509
	co-occurring in text	2367	689
	%co-occurring in text	18.90%	5.50%
	Mean co-occurrences per pair	1.2	0.36
Non-interacting gene pairs	Total	10,662,001	10,662,001
	co-occurring in text	72137	5019
	%co-occurring in text	0.68%	0.05%
	Mean co-occurrences per pair	0.011	0.0012
	%co-occurring pairs that are interacting pairs	3.28%	13.73%
	$R_{n>0}$	27.8	117.23

in sentences. For this reason, abstract co-occurrence is more sensitive but less specific for interactions. Of all pairs of genes co-occurring in text, only 3.3% of the pairs have physically interacting protein products. The remaining 96.7% might correspond to other types of interactions between proteins or genes, or might not indicate an interaction but rather just a vague relationship. So of any individual instance of a gene co-occurrence we might identify in an abstract, there is about a 1/30 chance that it corresponds to an actual physical interaction. On the other hand given any two random genes there is only a 0.12% chance that it corresponds to a physical interaction. So identifying a single instance of two genes co-occurring greatly increases the chances that it is in fact a true interaction. In fact to do a simple calculation:

$$\frac{P(\text{interact}|\text{co-occur})}{P(\text{interact})} = \frac{3.3\%}{0.12\%} = 27.5$$

So the simple presence of an abstract co-occurrence increases the probability that a pair of genes interacts by more than 20-fold.

About 19% of the 12,509 documented interactions co-occur among published abstracts; so abstract co-occurrence is only 19% sensitive for protein–protein interactions. However, only 0.7% of the 10,662,001 pairs of genes that are not documented interactions co-occur among the published abstracts; this corresponds to a specificity of 99.3%. This still corresponds to a huge number of false positives, however.

We can increase the specificity of interaction prediction at the cost of sensitivity by looking at co-occurrences of genes in sentences instead of abstracts. This may be a better strategy to mine the literature for protein–protein interactions, but a poorer one to verify predicted interactions. Of these co-occurring pairs, 13.7% of genes pairs have physically interacting protein products. The majority of the remaining probably corresponds to some sort of definite relationship between the genes. So for any individual pair of co-occurring genes in a sentence, there is about a 1/7 chance that it corresponds to an interaction in the GRID dataset. Repeating the above calculation:

$$\frac{P(\text{interact}|\text{co-occur})}{P(\text{interact})} = \frac{13.7\%}{0.12\%} = 114$$

The presence of a sentence co-occurrence increases the likelihood of an interaction by over 100-fold.

Only about 5.5% of the 12,509 documented interactions co-occur in sentences; this is considerably less sensitive than co-occurrence among abstracts. Only 0.05% of gene pairs that are not interactions co-occur in sentences; this corresponds to a specificity of 99.95%.

These results suggest that simple co-occurrence can offer an excellent opportunity to verify predicted interactions, assuming that the physical assay that predicts the interaction has a reasonable predictive value. The predictive value of an assay is the probability that a predicted interaction is a true interaction; it is similar to the precision of a prediction. Since co-occurrence among abstracts and sentences greatly increases the probability that a given pair of genes have a true physical interaction, the combination of a suggestive experiment and a co-occurrence ensure that the predicted interaction is in fact a true interaction. In the next section we will discuss how greater numbers of co-occurrences can increase the chance that a predicted interaction is a true interaction.

On the other hand, more than 80% of documented interactions have no co-occurrences in the text, even among abstracts. So, the presence of a co-occurrence can support the possibility of an interaction, but the lack of co-occurrence does not detract from the prediction made by the assay. Since abstract co-occurrences are only 19% sensitive for interactions, this method can help us only rarely. In this example we are looking at all genes, and not just well-studied genes. If we use this method on only well-studied genes, we can increase the sensitivity.

It is difficult to use only gene co-occurrence to predict gene networks accurately from text. Two genes co-occurring in the void of any additional experimental evidence does not necessarily provide strong evidence of an interaction. Even if we look at pairs of genes that co-occur in abstracts more than 32 times, 96 are protein–protein interactions while 160 are not. In fact, only 25% of the 2551 pairs that co-occur in abstracts more than eight times are protein–protein interactions. There is likely some sort of relationship between these other pairs – but we cannot be certain of the type of relationship. So it can be difficult to build accurate networks from the text using only co-occurrence.

They do provide a starting point for mining networks from text. For one thing, while co-occurrence implies a direct protein–protein interaction only rarely, it likely does imply other types of interactions between genes. In addition, it provides the greatest sensitivity in terms of what the text can offer. Other strategies for learning networks of protein–protein networks will use text analysis to only further filter and reduce the number of possible interactions. Abstract co-occurrence casts the widest net of text-based strategies. Since only 19% of all protein–protein interactions are picked up by co-occurrences, the majority of known interactions are actually probably not documented in the text by any discernible means.

Looking at sentence co-occurrences provides much stronger evidence; it is much more specific for protein–protein physical interactions. However, the sensitivity is less – fewer than 6% of the documented interactions are picked up by sentence co-occurrences, and even fewer have multiple co-occurrences. So, while fewer true interactions are picked up by sentence co-occurrence, the interactions are more likely to correspond to true protein–protein interactions.

10.6 Number of textual co-occurrences predicts likelihood of an experimentally predicted interaction

Let us suppose that we have an experimental result that suggests that two genes interact. Let us then suppose that a gene pair is observed to co-occur in a certain number of abstracts. We can calculate the probability of the predicted interaction being a true interaction based on the number of co-occurrences. In this section we assume that the experimental result suggests a prior probability of interaction, $P(\text{interact})$. This is the predictive value or precision of the experiment. For many of the high throughput assays, P(interact) might be considerably less than 1. The probability of the interaction with the additional information of the number of co-occurrences in text can be computed. By Bayes' theorem (see Section 2.2) we can calculate:

$$P(\text{interact}|n \text{ co-occur}) = \frac{P(n \text{ co-occur}|\text{interact})P(\text{interact})}{P(n \text{ co-occur})}$$

We can expand this to:

$$\frac{P(n \text{ co-occur}|\text{interact})P(\text{interact})}{P(n \text{ co-occur}|\text{interact})P(\text{interact})+P(n \text{ co-occur}|\text{non-interact})P(\text{non-interact})}$$

We define an entity R_n

$$R_n = \frac{P(n \text{ co-occur}|\text{interact})}{P(n \text{ co-occur}|\text{non-interact})}$$

This is the ratio of the probability that n co-occurrences occur given that two genes interact to the probability that the same number of co-occurrences occur given two genes don't interact. The parameter R is a measure of the predictive value of that number of co-occurrences in the text.

Then we can simplify:

$$P(\text{interact}|n \text{ co-occur}) = \frac{R_n P(\text{interact})}{1 + (R_n - 1)P(\text{interact})}$$

So as the term R_n gets larger and larger, the probability that the two genes interact approaches 1. Of course, if the prior probability is extremely small, and the product of R_n and P(interact) is a small number, then the probability of the predicted interaction being real is also small. One case when P(interact) might be very small is when there is absolutely no available experimental evidence for a particular interaction.

In Table 10.2 we have listed the values of $R_{n>0}$. Given one or more abstract co-occurences, the R value is 28. So if abstract co-occurrences are documented and an assay predicts that interaction with a predictive value of at least 1/7, then the probability of a true interaction is at least 4/5. In other words, even a very modestly effective assay combined with any degree of abstract co-occurrence is quite a strong predictor of a true interaction.

One would expect that as the number of abstract co-occurrences for a pair of genes increases, the likelihood of an interaction increases as well. In other words, R_n should be proportional to n. In Figure 10.1 we have plotted the probability that two genes co-occur n times given that the two genes are

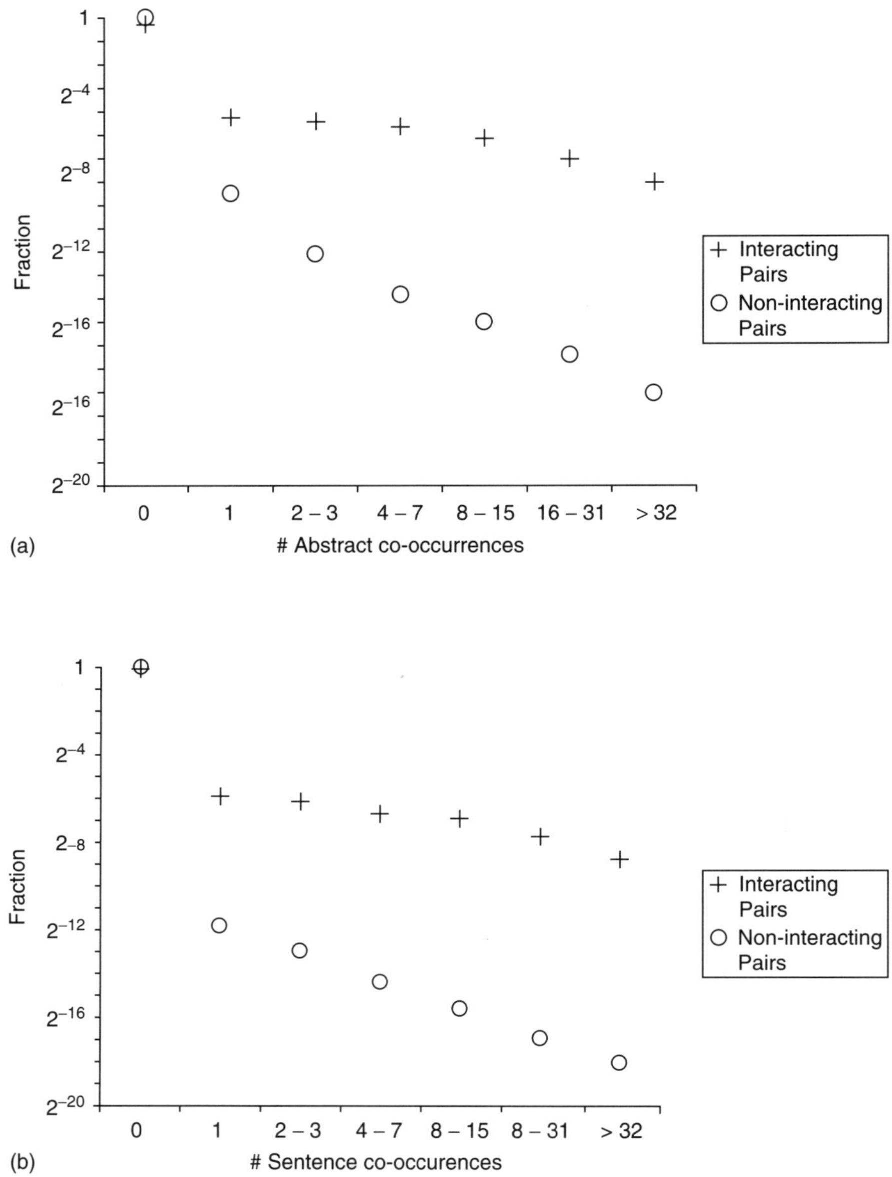

Figure 10.1 *Probability of n co-occurrences in text.* In both plots pluses (+) indicate the probability of the number of co-occurrences on the x-axis if the co-occurring genes were known to interact. In both plots circles (o) indicate the probability of the number of co-occurrences on the x-axis if the co-occurring genes were known not to interact. (a) For abstract co-occurrences. (b) For sentence co-occurrences.

known to interact, and also the probability that they co-occur n times if they are known not to interact. In Figure 10.2 we plot R_n, the ratio of those probabilities. While for a single abstract co-occurrence R is only about 10, for 4 to 7 co-occurences R is about 150, and for greater than 32 abstract co-occurrences R is 511. It is evident that as the number of abstract co-occurrences increase, the chance that a predicted interaction is a true interaction also increases dramatically. In Figure 10.3(a) we have plotted the necessary prior probability to obtain specific interaction probabilities for different numbers of abstract co-occurrences. The y-axis indicates the prior probability, the x-axis indicates the number of co-occurrences in text. Each line corresponds to a specific probability of interaction; for example, each point on the top line corresponds to a prior probability and the number of co-occurrences necessary to achieve an interaction probability of 0.9. From this graph we can see the dramatic affect that the number of co-occurrences can have on the probability of an interaction. An assay that suggests a protein interaction with a prior probability of 0.01 and has one co-occurrence has an interaction probability of 0.1. On the other hand, if 8–15 co-occurrences are observed then the probability of interaction becomes 0.7.

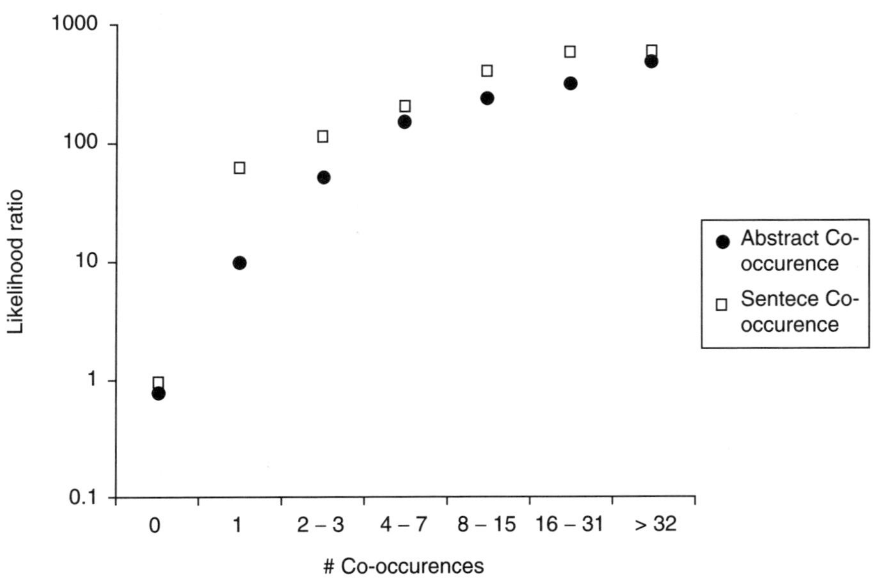

Figure 10.2 *Plot of R as a function of number of co-occurrences.* Here, for a given number of co-occurrences (x-axis), we have plotted the ratio of the probability of that number of co-occurrences assuming that the pair interacts to the probability assuming that the pair does not interact. We refer to this ratio as R. Squares (□) represent sentence co-occurrences; dark circles (●) represent abstract co-occurrences.

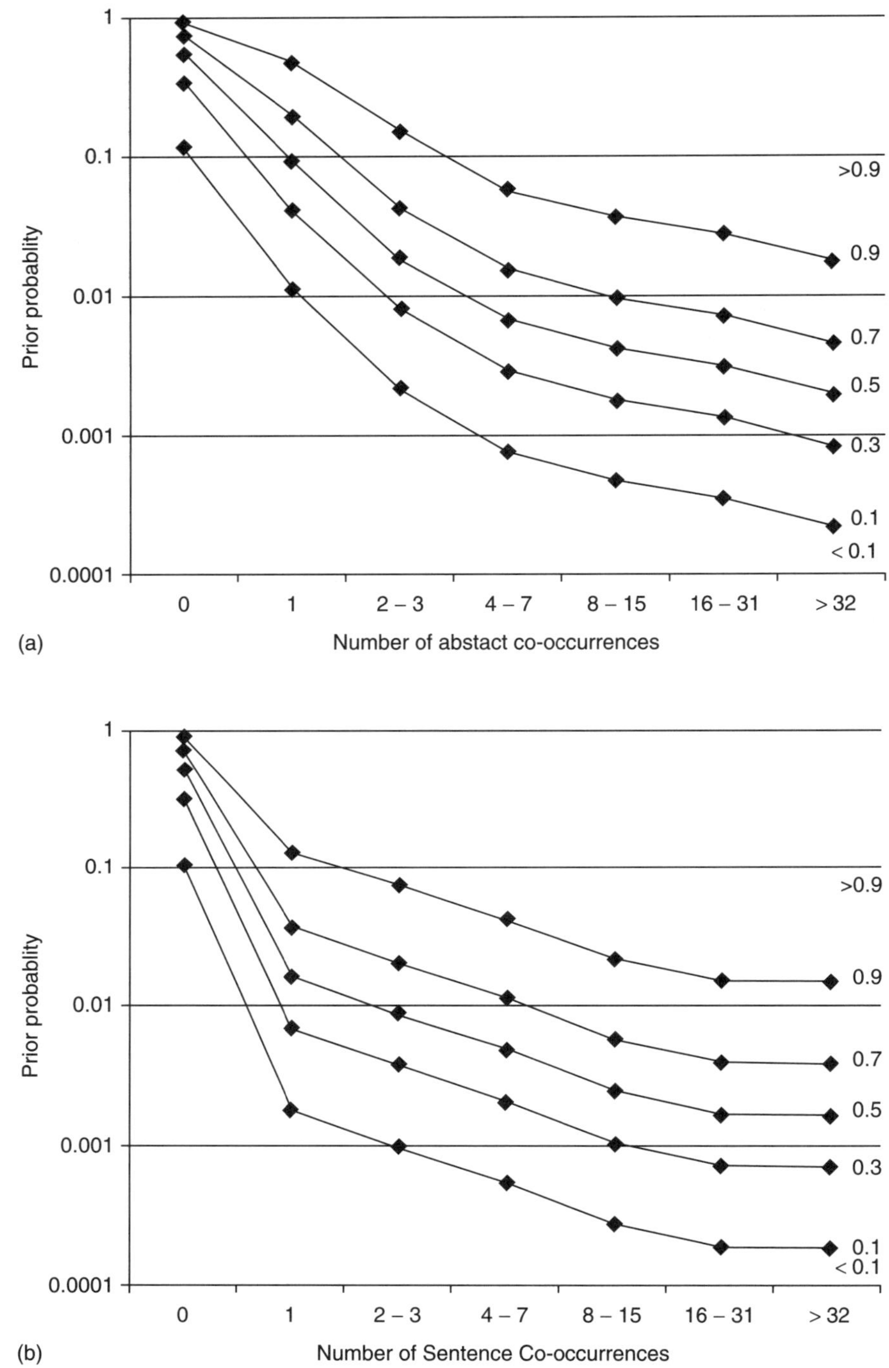

Figure 10.3 *Relationship between the number of co-occurrences in the text, the prior probability of an interaction, and the ultimate probability of the interaction.* On the x-axis we plot the number of co-occurrences in the literature. On the y-axis we plot the prior probability on an interaction. Typically that value is based on an experimental assay such as yeast-2-hybrid.

Compared to abstract co-occurrences, the R-values are greater for each n for sentence co-occurrences (see Figure 10.2). So a single sentence co-occurrence has an R of 61 compared to 9 for a single abstract co-occurrence. Similarly 8–15 sentence co-occurrence has an R of 406 compared to 235 for 8–15 abstract co-occurrences. The R value for all $n > 0$ is 117. In Figure 10.3(b) we have plotted the necessary prior probability to obtain specific interaction probabilities for different numbers of sentence co-occurrences. Notice that the curves are shifted down compared to Figure 10.3(a), indicating that lower prior probabilities are necessary to obtain the same interaction probability.

In both of these cases it is very apparent that as the number of co-occurrences increases, the probability of an interaction quickly increases.

10.7 Information extraction and genetic networks: increasing specificity and identifying interaction type

As is apparent from the analysis on co-occurrence, the mere presence of two gene names in the same abstract or even the same sentence provides supportive evidence of an interaction, but certainly offers only minimally suggestive evidence of the possibility of a protein–protein interaction in the void of experimental data. The difficulty is that the presence of two genes in the same sentence may imply many different sorts of relationships besides just a protein–protein interaction. To build text-based genetic networks we need more specific evidence than just a co-occurrence. In the remainder of this chapter we will discuss strategies to select sentences that focus on protein interactions specifically. These strategies will hopefully eliminate the sentences with gene co-occurrences that do not address protein interactions.

Information extraction is a more focused approach to mining the text. The goal of information extraction is to mine specific information from the text. Typically this is achieved by identifying specific patterns in the text. These patterns might include part of speech, sentence structure, and the presence of keywords in the sentence.

Figure 10.3 (*Caption Continued*)
The lines represent fixed probabilities of an interaction given a prior probability and a number of co-occurrences. Given a fixed prior probability, and number of observed co-occurrences we can use this plot to determine the probability of the interaction. (a) Abstract co-occurrences. For example if there are more than four abstract co-occurrences and the prior probability of an interaction based on an experimental assay is 0.1, then the true probability of an interaction is > 0.9. (b) Sentence co-occurrences. For example if there are more than two sentence co-occurrences and the prior probability of an interaction based on an experimental assay is 0.1, then the true probability of an interaction is > 0.9.

One very effective strategy to identify sentences describing true protein–protein interactions is to look at sentences where gene names co-occur with protein–protein interaction keywords in the sentence. This should eliminate many of the erroneous co-occurrences. Blaschke and colleagues proposed a set of high quality verbs that are commonly used to describe protein–protein interactions (Blaschke, Andrade et al. 1999). We have listed these verbs in Table 10.3. This same group looked for a specific pattern in the text to identify potential interactions; they looked for text that contained two gene names and one of these key verbs in between them. This is an example of information extraction. In this case we are looking for a simple pattern:

Gene A . . . *interaction verb* . . . Gene B

If this pattern is observed by the algorithm, it reports that Gene A interacts with Gene B via the mechanism indicated by the *interaction verb*.

They used these text segments to predict the presence of a protein–protein interaction between these two genes. They were able to successfully and accurately reconstruct specific small networks of genes in *Drosophila*. They also used the verb keywords to predict the type of interaction based on the keyword between the gene names (for example, binds versus phosphorylates). This is a very effective strategy to filter co-occurrences.

To demonstrate the effectiveness of this approach, we conducted a simple test. We selected 200 random sentences with co-occurring gene names. We manually examined these sentences and determined that a total of 83 (41.5%) could support or imply a physical interaction between the two proteins in the sentences. Using the template proposed here to filter sentences, we found that a total of 58 sentences were selected, of which 36 could support or imply a physical interaction. So using this template to select sentences reduced the recall from 100% to 43% (36 out of 83). On the other hand, it increases the precision from 41.5% to 62% (36 out of

Table 10.3 *Words identified by Blaschke and colleagues to identify protein interactions.* These were the words that were used to look for protein interaction patterns in text. The left-most word represents the root. The suffixes on the right represent possible modifications of that root.

acetylat-	-e	-ed	-es	-ion
activat-	-e	-ed	-es	-ion
associated with bind	-ing	-s	-s to	/bound
destabiliz	-e	-ed	-es	-ation
inhibit	-ed	-es	-ion	
interact	-ed	-ing	-s	-ion
is conjugated to modulat	-e	-ed	-es	-ion
phosphorylat	-e	-es	-es	-ion
regulat	-e	-es	-es	-ion
stabiliz	-e	-ed	-es	-ation
suppress	-ed	-es	-ion	
target				

58). So while filtering sentences with co-occurring gene names for specific patterns can cause a certain loss of sensitivity, the selected sentences are in fact much more predictive of a true gene interaction.

We apply this approach to all of the sentences with gene name co-occurrences and correlate them with the GRID interaction data set. The results are summarized in Table 10.4. Notice that many sentences are eliminated when we accept only those that conform to Blaschke's pattern. Instead of 17,712 co-occurrences among 5708 pairs of genes, we have only 3891 co-occurrences among 1969 pairs of genes. This approach is much less sensitive, since sentences describing or suggesting interactions could be filtered out. Only 3% of the GRID interacting pairs are represented among these filtered sentences. A full 45% of the interacting gene pairs identified when we included all sentences with co-occurring names are lost when we apply this pattern-matching filter. However, these sentences are a much more specific set. The parameter R is greater than 200, suggesting greater specificity. Also 19% of the total co-occurrences correspond to interacting gene pairs; this is an increase from 13% when considering all sentences with co-occurrences.

Table 10.4 *Data for interactions predicted by sentences that co-occur and contain patterns suggestive of potential interactions.* This table is similar Table 10.2. However instead of just co-occurrences in sentence, we look for sentences with co-occurring gene names that conform to a pattern described by Blashke and colleagues. These co-occurrences are more likely to describe true protein interactions.

		Sentences matching patterns
Text	co-occurrences	3891
	Pairs of co-occurring genes	1969
	%co-occurrences that correspond to interacting genes	30.27%
Interacting gene pairs	Total	12,509
	co-occurring in text	377
	%co-occurring in text	3.01%
	Mean co-occurrences per pair	0.094
Non-interacting gene pairs	Total	10,662,001
	co-occurring in text	1592
	%co-occurring in text	0.014%
	Mean co-occurrences per pair	0.00025
	%co-occurring pairs that are interacting pairs	19.15%
	$R_{n>0}$	201.84

Information extraction can also utilize patterns in part of speech as well as word usage. Other groups showed how part of speech can be used in addition to keywords to identify protein–protein interactions (Sekimizu, Park et al. 1998; Thomas, Milward et al. 2000; Ono, Hishigaki et al. 2001). First, part of speech is assigned to sentences with two gene names. Then specific patterns including part of speech features are identified that might correspond to potential interactions. The methods can be complicated since each type of interaction requires a definition of many different patterns including different arrangements of part of speech and different variants of the same interaction verb.

An even more sophisticated approach is to parse sentences, and identify the subject and noun phrases. If the phrases are gene names, then the next step is to identify the predicate of the sentences, and assess if it is consistent with an interaction. This approach was proposed and preliminarily evaluated by Sekimizu and colleagues. (Sekimizu, Park et al. 1998). While these methods that involve looking for complicated specific patterns in text can greatly reduce the sensitivity, they can be much more predictive of true protein–protein interactions. They are better suited for identifying protein interactions in the void of experimental evidence.

10.8 Statistical machine learning

Statistical methods can also be used to analyze sentences to identify those that are strongly suggestive of interactions. Statistical textual classifiers can be used as an alternative to rule-based information extraction methods to identify protein–protein interactions. In Chapter 8 we demonstrated how classifiers are used to find text that describes specific biological function. The same classifiers can be used to identify segments of text that are potentially describing an interaction based on word usage.

One group proposed using a naive Bayes classifier to determining whether an abstract is relevant to a protein–protein interaction (Marcotte, Xenarios et al. 2001). Naive Bayes is detailed in chapter 8. They trained their classifier on documents known to describe protein–protein interactions. They then used the classifier to scan unclassified abstracts and determine if they are potential references that are describing protein–protein interactions. This group devised this approach to assist in augmenting the reference base of the Database of Interacting Proteins (DIP), a well-known repository of documented protein–protein interactions.

Another similar application of text classification was implemented to identify interactions described in the text for the Biomolecular Interaction Network Database (BIND) (Donaldson, Martin et al. 2003). Instead of naive Bayes, this group implemented support vector machine text classifiers. They demonstrated that support vector machine classifiers can be much more effective than naive Bayes in classifying abstracts relevant to protein–protein interactions. In one test it achieved more than 12% more recall. They then demonstrated that they could use the same classifier to identify critical sentences in the abstract that were relevant to protein interactions. They predicted the presence of protein interactions by looking for sentences that contained two gene names and that received a high score by the support vector machine classifier. They estimated that this approach identified about 60% of known interactions in yeast. In addition they found that automated methods saved 70% of the manual time spent to identify protein interactions from text.

As an example, we demonstrate statistical classification of those sentences with co-occurring gene names. We use a maximum entropy textual classifier to distinguish sentences that describe protein interactions from others. The technical aspects of the classifier are described in Chapter 8. We select 500 random abstracts from the January 2002 release of DIP as a positive training set. We also select 500 random yeast references as a negative training set. Words appearing in more than four but fewer than 500 documents were selected. Chi-squared testing is used to select 2000 of those vocabulary words that differentiated these two sets. For each of the two sets, maximum entropy features were defined for each of these vocabulary words. A maximum entropy classifier is trained on these two document sets. We then use the classifier to predict whether a sentence is more consistent with the positive or negative set. Sentences are assigned probabilities that are presumably proportional to the confidence with which the sentence was in fact describing an interaction.

Probability thresholds can be selected to achieve different degrees of precision and recall. We examine the set of 200 manually analyzed sentences described in the previous section. We had determined that 83 of those sentences suggested protein interactions. If we apply the maximum entropy classifier to these sentences, we find that a total of 104 sentences have a probability greater than 0.5; of these 104 sentences, 65 describe interacting proteins per our manual analysis. This corresponds to a precision of 62% (65 out of 104) and a recall of 78% (65 out of 83 manually selected sentences). We have listed selected examples of these sentences in Table 10.5. For each sentence we have listed the probability assigned by maximum entropy that the sentence represents a protein–protein interaction and also the objective assignment of whether the sentence represents an interaction. The six very high scoring sentences unambiguously

Table 10.5 *Example of sentences with two gene names, and their probability of describing a protein–protein interaction.* In this table we list some examples of randomly selected sentences. The gene names have been replaced with the formal yeast gene name within brackets. In the first column we list the probabilistic score assigned by the maximum entropy classifier. The second column lists the expert assessment of the sentence; the number 1 indicates that the expert thought the sentence was describing or might suggest a protein–protein interaction. The final column lists the sentence in question. There are three groups of sentences. The six highest scoring sentences are listed at the top. These sentences unambiguously describe protein–protein interactions. Then we list four sentences that received intermediate probabilities. Finally we list four sentences that are low probability for describing protein–protein interactions.

ME probability	Expert score	Sentence
0.93	1	Gel filtration of the <YNL262W> p. <YPR175W> p complexes reveals a novel heterotetrameric form, consisting of two heterodimers of <YNL262W> p. <YPR175W> p.
0.93	1	Thus the <YCL032W> p- <YLR362W> p interaction may differentially modulate the flow of information through the various mapk-mediated pathways.
0.92	1	<YOR181W> protein is localized to actin patches and interacts with <YBL007C> p, a src homology 3 domain-containing protein previously implicated in actin assembly and function.
0.90	1	<YBR112C> (<YBR112C>)- <YCR084C>, a general co-repressor complex, is recruited to promoter dna via interactions with dna-binding regulatory proteins and inhibits the transcription of many different yeast genes.
0.90	1	The central receptor <YNL131W> binds preproteins through both its cytosolic domain and its intermembrane space domain and is stably associated with the channel protein <YMR203W> (refs 11-13).
0.88	1	<YFL029C> binds tightly to and phosphorylates <YBR160W>, thereby allowing its subsequent activation by the binding of a cyclin.
0.52	0	These data indicate that the degradation of <YAL040C> involves <YBR160W> -dependent phosphorylation events.
0.51	1	The last 115 amino acids of <YDR390C>, which contains an 82- amino acid region not present in previously characterized e1 enzymes, is sufficient for the interaction with <YKR002W>.
0.51	1	Immunofluorescent staining with anti-human <YOL069W> p and with anti-hec, the human homologue of <YIL144W> p, showed that both proteins are at the centromeres of mitotic hela cells.
0.50	0	Another gene, <YBR133C>, is a novel negative regulator of <YJL187C> function.
0.14	0	<YFL029C> was previously identified as a multicopy suppressor of a weakened <YPR054W> mutant and shown to be required for spore wall assembly.

Table 10.5 *Continued*

ME probability	Expert score	Sentence
0.13	0	Transcription activation of sulfur metabolism in yeast is dependent on two dna binding factors, the centromere binding factor 1 (<YJR060W>) and <YNL103W>.
0.06	0	This, together with reciprocal genetic interactions between <YER133W> and <YKL193C>, suggests that <YKL193C> p functions positively with <YER133W> p to promote dephosphorylation of nuclear substrates required for faithful transmission of chromosomes during mitosis, and this role is at least partly mediated by effects of <YKL193C> p on the nuclear distribution of <YER133W> p
0.04	0	The essential ras-related gtpases <YFL038C> and <YFL005W> act at distinct stages of the secretion pathway in the yeast saccharomyces cerevisiae: <YFL038C> is required for vesicular transport from the endoplasmic reticulum to the golgi apparatus, whereas <YFL005W> is required for fusion of secretory vesicles to the plasma membrane.

represent protein–protein interactions. The four sentences listed with intermediate probabilities do suggest possible protein–protein interactions, but are not definitive. For example, the first sentence says that one protein relies on phosphorylation events from another protein – presumably that represents a direct interaction, though the sentence does not make it unambiguously clear that the one protein is phosphorylating the other. The final four sentences are low probability sentences; it is clear they represent no physical protein–protein interation.

For different thresholds we have plotted precision as a function of recall in Figure 10.4. A threshold can be selected that achieves the appropriate level of precision and recall. Notice that for the same level of recall as the method described by Blaschke, maximum entropy classification obtains a precision greater than 80%; at that threshold less than 1 in 5 selected sentences do not describe protein–protein interactions.

We can also see from this data that the higher the probability assigned to a sentence by maximum entropy classification, the more likely it is actually describing an interaction. This is demonstrated in Figure 10.5. In this figure we have partitioned the 200 sentences by the probability confidence value assigned to them. Notice that the fraction of genes suggesting a true protein interaction tracks with the confidence scores.

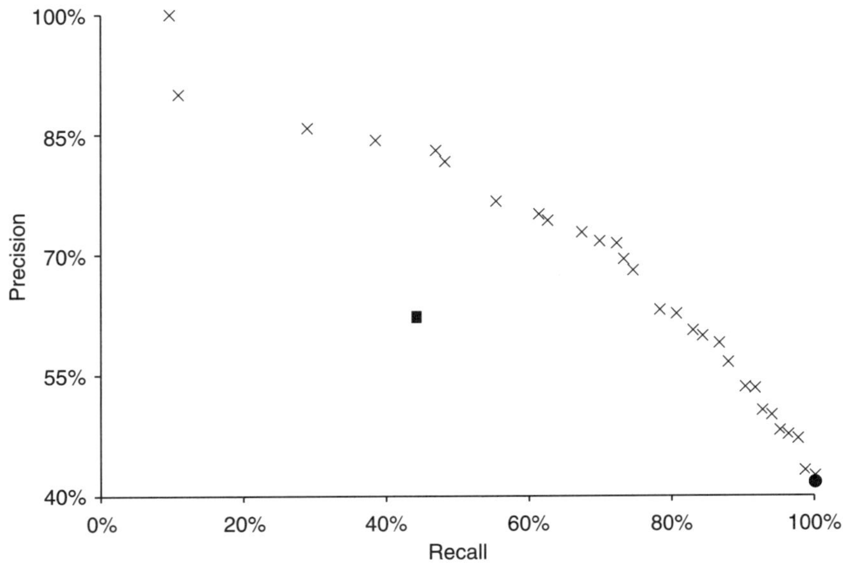

Figure 10.4 *Precision–Recall plot for maximum entropy classification of sentences with co-occurring genes.* Here we use maximum entropy classification to determine which sentences are describing protein interactions. The dark circle represents the precision and recall for simply taking all sentences with co-occurring gene names. The dark square represents the precision and recall for the strategy proposed by Blaschke and colleagues in this set. The cross symbols represent the precision and recall of maximum entropy classification of the sentences. Notice that as we apply more stringent criteria, the recall is reduced since we are losing some of the sentences that describe protein–protein interactions. On the other hand, as we apply increasingly stringent criteria the precision increases since sentences describing other phenomena are eliminated. We can be as stringent as necessary to obtain the necessary precision.

We can use maximum entropy to classify all of the sentences with co-occurring protein names. The selected sentences should describe true protein interactions. In Table 10.6 we list our results for a threshold of 0.5 and 0.7. If we use a probability threshold 0.5 to distinguish sentences describing interactions, we find that the total number of sentences with co-occurring genes is reduced from 17,712 to 6498. Of these, 36% of the sentences contain co-occurring proteins that the GRID data set lists as interacting proteins. These sentences with co-occurring gene names correspond to 2711 gene pairs, of which 21% are interacting gene pairs. A total of 4.5% of the GRID interactions are represented among these sentences. This represents only a loss of less than 17% of the interacting gene pairs obtained when using all sentences with co-occurring gene names. Using a maximum entropy classifier with a 0.5 cutoff to screen sentences identifies 52% more gene pairs than Blaschke's approach. So it is a more sensitive approach. It also has a slightly greater predictive power; it achieves an R value of 229. This is also reflected in the fact that a greater percentage of selected sentences correspond to true protein interactions.

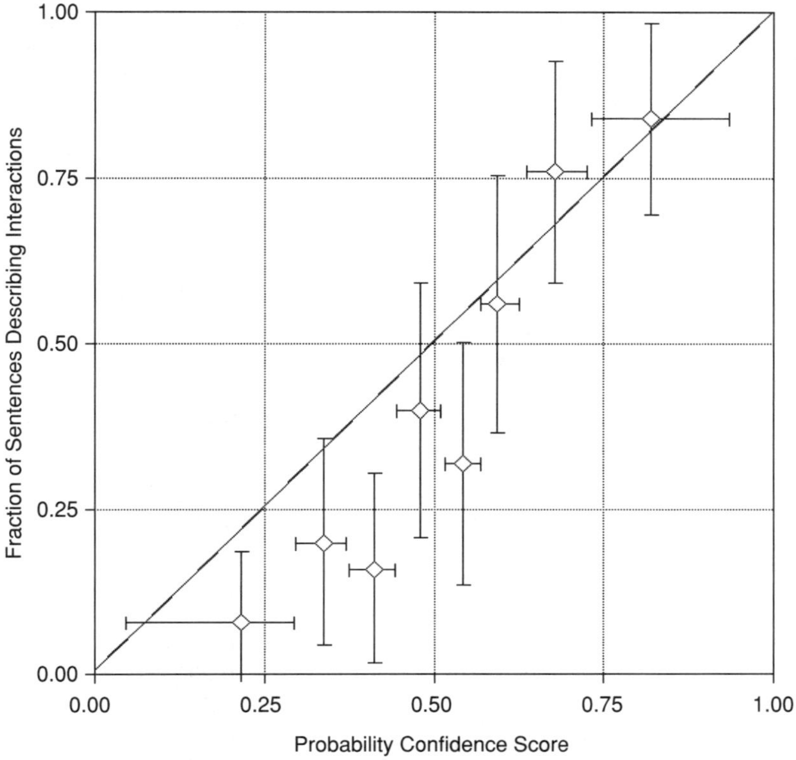

Figure 10.5 *Sentences suggesting protein–protein interactions as a function of maximum entropy confidence scores.* The *y*-axis in this plot represents the fraction of sentences that have been manually assessed to suggest protein interactions. Sentences were classified with the maximum entropy classifier and binned according to confidence score; that is the classifier probability that the sentence represents a protein–protein interaction. The fraction of those sentences that actually describe interactions is proportional to the confidence score.

Increasing the probability threshold to 0.7 decreases the recall, but greatly increases the specificity of the method. The 17,712 sentences with gene co-occurrences is reduced to 1575 sentences, of which 43% correspond to interacting proteins. Only 2.6% of the interacting gene pairs in GRID are obtained from these sentences.

Prediction of protein interactions could be made even more accurate by combining the probability scores for different sentences that mention the same pairs of genes. Pairs that have multiple high probability sentences almost certainly interact with each other.

We also remind the reader that these methods may be more precise than suggested by the comparison to the GRID data set, since the GRID data set does not include all of the known interactions by any means. So many of the predicted interactions not supported by GRID are true interactions. Effective

Table 10.6 *Data for interactions predicted by sentences selected by maximum entropy calssification.* This table is similar Table 10.2. However instead of just co-occurrences in sentences, we look for sentences selected by the maximum entropy classifier described in the text. These co-occurrences are more likely to describe true protein interactions.

		$p > 0.5$	$p > 0.7$
Text	co-occurrences	6493	1775
	Pairs of co-occurring genes	2711	998
	%co-occurrences that correspond to interacting genes	36.04%	43.11%
Interacting gene pairs	Total	12,509	12,509
	co-occurring in text	575	328
	%co-occurring in text	4.59%	2.62%
	Mean co-occurrences per pair	0.187	0.054
Non-interacting gene pairs	Total	10,662,001	10,662,001
	co-occurring in text	2136	670
	%co-occurring in text	0.02%	0.0062%
	Mean co-occurrences per pair	0.00039	0.000084
	% co-occurring pairs that are interacting pairs	21.21%	32.87%
	$R_{n>0}$	229.45	417.27

use of statistical text classification starts to approach the level of accuracy that is appropriate to start building high quality networks of interacting genes.

We also note that while the focus of this chapter has been protein–protein interactions, all of these same principles can be applied to other types of interactions between genes and their products. Statistical text classifiers can easily be constructed to identify specific types of protein–protein interactions or other types of interactions in sentences with gene name co-occurrences.

References

Blaschke, C., M. A. Andrade, et al. (1999). "Automatic extraction of biological information from scientific text: protein-protein interactions." *Proc. Int. Conf. Intell. Syst. Mol. Biol.* **2**(1): 60–7.

Breitkreutz, B. J., C. Stark, et al. (2003). "The GRID: the General Repository for Interaction Datasets." *Genome Biol.* **4**(3): R23.

Donaldson, I., J. Martin, et al. (2003). "PreBIND and Textomy – mining the biomedical literature for protein-protein interactions using a support vector machine." *BMC Bioinformatics* **4**(1): 11.

Gavin, A. C., M. Bosche, et al. (2002). "Functional organization of the yeast proteome by systematic analysis of protein complexes." *Nature* **415**(6868): 141–7.

Giot, L., J. S. Bader, et al. (2003). "A protein interaction map of Drosophila melanogaster." *Science* **302**(5651): 1727–36.
Ho, Y., A. Gruhler, et al. (2002). "Systematic identification of protein complexes in Saccharomyces cerevisiae by mass spectrometry." *Nature* **415**(6868): 180–3.
Ito, T., T. Chiba, et al. (2001). "A comprehensive two-hybrid analysis to explore the yeast protein interactome." *Proc. Natl. Acad. Sci. U S A.* **98**(8): 4569–74.
Jenssen, T. K., A. Laegreid, et al. (2001). "A literature network of human genes for high-throughput analysis of gene expression." *Nat. Genet.* **28**(1): 21–8.
Marcotte, E. M., I. Xenarios, et al. (2001). "Mining literature for protein-protein interactions." *Bioinformatics* **17**(4): 359–363.
Ono, T., H. Hishigaki, et al. (2001). "Automated extraction of information on protein-protein interactions from the biological literature." *Bioinformatics* **17**(2): 155–61.
Rain, J. C., L. Selig, et al. (2001). "The protein–protein interaction map of Helicobacter pylori." *Nature* **409**(6817): 211–5.
Sekimizu, T., H. S. Park, et al. (1998). "Identifying the interaction between genes and gene products based on frequently seen verbs in MedLine abstracts." *Genome Inform. Ser. Workshop Genome Inform.* **9**: 62–71.
Thomas, J., D. Milward, et al. (2000). "Automatic extraction of protein interactions from scientific abstracts." *Pac. Symp. Biocomput*: 541–52.
Uetz, P. (2002). "Two-hybrid arrays." *Curr. Opin. Chem. Biol.* **6**(1): 57–62.
Uetz, P., L. Giot, et al. (2000). "A comprehensive analysis of protein–protein interactions in Saccharomyces cerevisiae." *Nature* **403**(6770): 623–7.
Walhout, A. J., R. Sordella, et al. (2000). "Protein interaction mapping in C. elegans using proteins involved in vulval development." *Science* **287**(5450): 116–22.
Zhu, H., M. Bilgin, et al. (2003). "Proteomics." *Annu. Rev. Biochem.* **72**: 783–812.

11 Conclusion

The genomics era has presented many new high throughput experimental modalities that are capable of producing large amounts of data on comprehensive sets of genes. In time there will certainly be many more new techniques that explore new avenues in biology. In any case, textual analysis will be an important aspect of the analysis. The body of the peer-reviewed scientific text represents all of our accomplishments in biology, and it plays a critical role in hypothesizing and interpreting any data set. To altogether ignore it is tantamount to reinventing the wheel with each analysis.

The volume of relevant literature approaches proportions where it is all but impossible to manually search through all of it. Instead we must often rely on automated text mining methods to access the literature efficiently and effectively.

The methods we present in this book provide an introduction to the avenues that one can employ to include text in a meaningful way in the analysis of these functional genomics data sets. They serve as a complement to the statistical methods such as classification and clustering that are commonly employed to analyze data sets. We are hopeful that this book will serve to encourage the reader to utilize and further develop text mining in their own analyses.

INDEX

Note: Authors of cited works appear under the first cited name only. Page numbers in *italics* refer to figures, and those in **bold** to tables. The abbreviation, 'pl.' refers to plates.

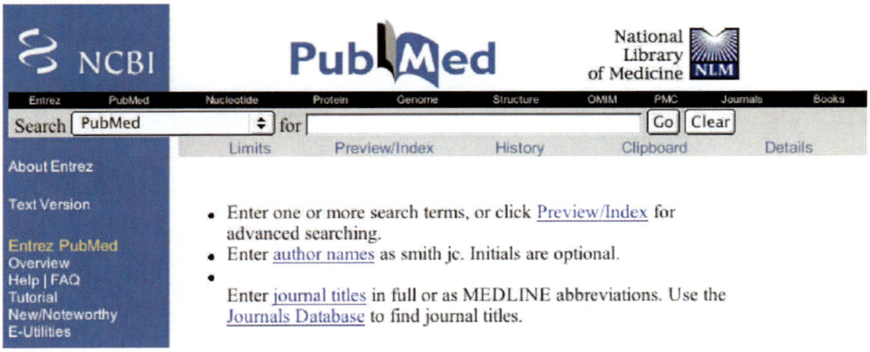

Plate 1.1 *The PubMed database homepage.* This is a screen shot of the PubMed home page. This interface is used to search PubMed's database of 14,000,000 articles references.

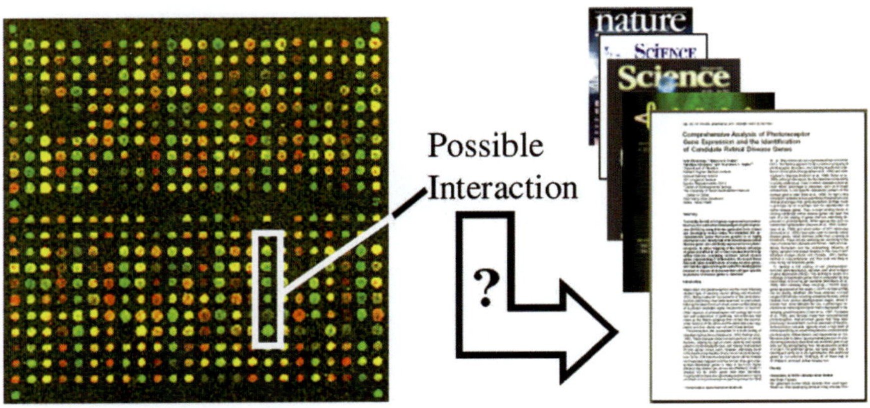

Plate 1.2 *Schematic of using text analysis in genomic data analysis.* In this schematic we have shown sample gene expression array data. Each spot represents the expression of a given gene by its color. The white box indicates a collection of genes that seem to have similar expression; this suggests the possibility of an interaction between these genes. This interaction must ultimately be verified by detailed experimentation. But an initial survey may include querying the scientific literature to see if there are clues that suggest a possible relationship in the text.

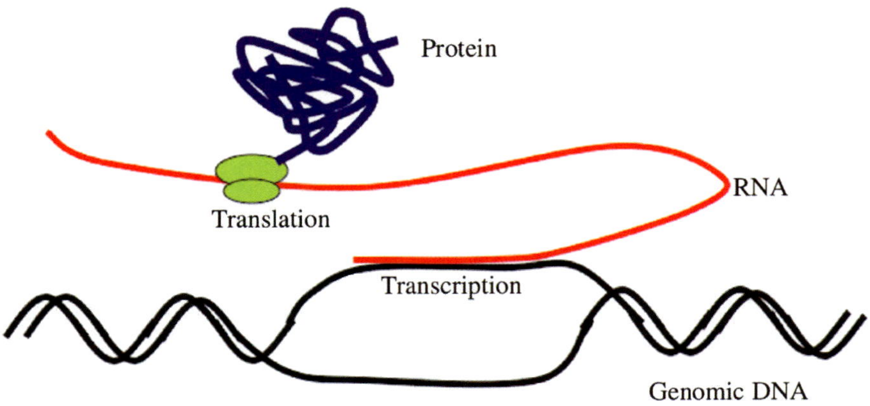

Plate 2.1 *The central dogma of biology.* Genomic DNA is the basic building block of life, and contains the blueprints for all of the proteins in the cell. It is reproduced and passed on to progeny cells. The double stranded helix of DNA can be opened up and used as a template by RNA polymerase to make messenger RNA. The ribosome (depicted in green) can read the messenger RNA and translate it into a linear polymer of amino acids, or a polypeptide. The polypeptide is folded into a protein, which can then be used to perform different cellular functions.

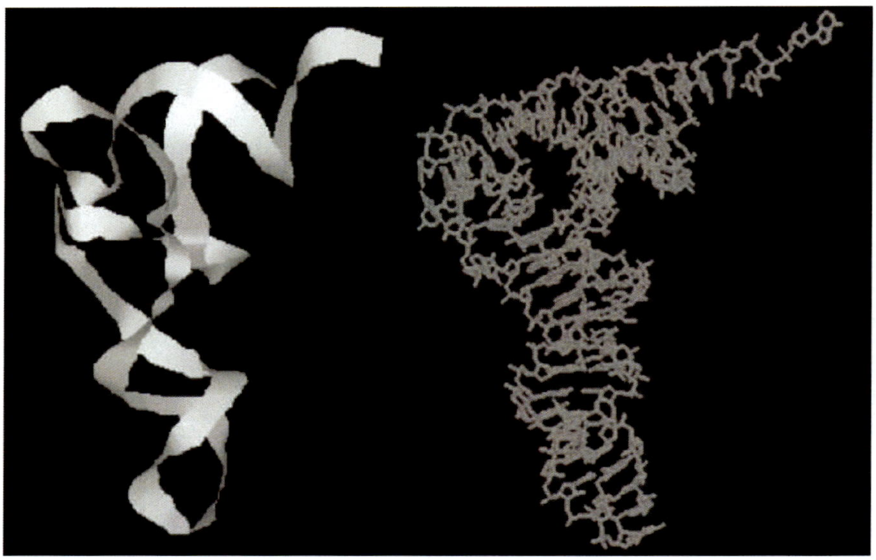

Plate 2.2 *Yeast phenylalanine tRNA.* Depicted here is the structure of a yeast transfer RNA. The information for this structure was obtained from the Protein Databank (1ehz). RNA folds in on itself to form a complicated three-dimensional structure that is suited for binding messenger RNA in the ribosome and attaching an appropriate amino acid to an elongating polypeptide chain. In this case, phenylalanine is the amino acid carried by this particular mRNA.

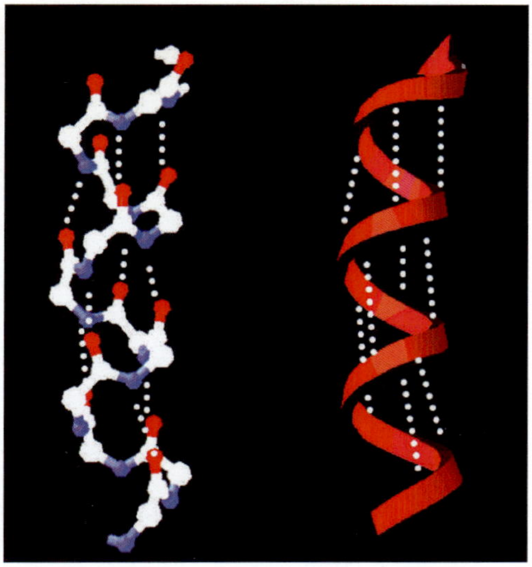

Plate 2.3 *Hydrogen bonding in alpha helices.* In this figure the hydrogen bonding that stabilizes alpha helices is illustrated. On the left the backbone of a peptide chain that has folded into an alpha helix is depicted. On the right the same structure is depicted in ribbon form for easy viewing. The red atoms represent oxygen, white atoms represent carbon, and blue atoms represent nitrogen from the amino groups. Hydrogen bonding is between carboxyl group oxygen and the hydrogen from the amino group. In this depiction the N-terminal is at the bottom of the figure and the C-terminal is at the top. All carboxyl groups point upwards, and the amino groups point downward; they interact by non-covalent hydrogen bonding and stabilize this structure. The amino acid side chains typically point outward from the helix; they, along with the hydrogen atoms, are not displayed here.

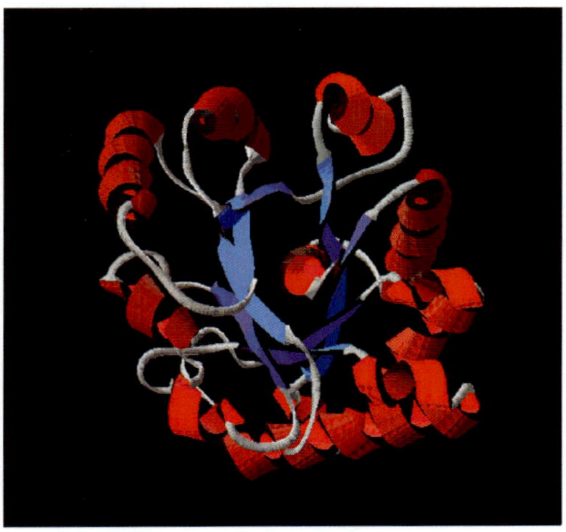

Plate 2.4 *Structure of triose phosphate isomerase.* Ribbon diagram for triose phosphate isomerase. This structure consists of eight alpha helices surrounding eight parallel beta sheets. This canonical structure is a common one in nature, found most often to have an enzymatic function. Triose phosphate isomerase is a key enzyme involved in glucose metabolism. The structure was obtained from the Protein Databank (1amk).

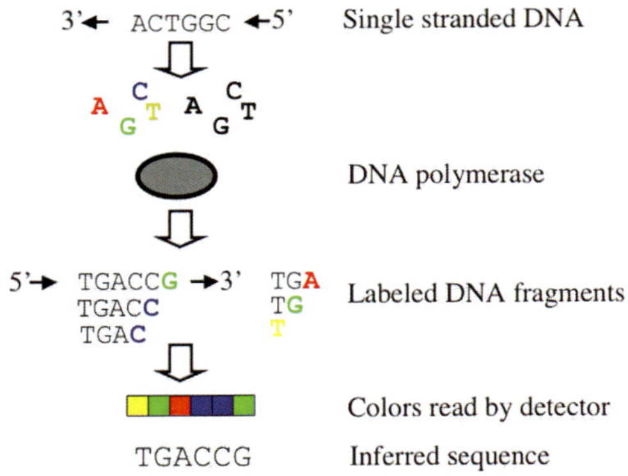

Plate 2.5 *The Sanger sequencing method.* Single stranded DNA is isolated. The single strand is used as a template for DNA polymerase. Normal nucleotides are used as well as fluorescently labeled chain terminating nucleotides to synthesize new DNA fragments. Each of the four types of nucleotides is labeled with a different colored dye. Each fragment is labeled only with the color corresponding to the nucleotide they terminate in. Then the fragments are sorted by length, and then their colors are recognized by a detector. The sequence of colors corresponds to the DNA sequence of the original strand.

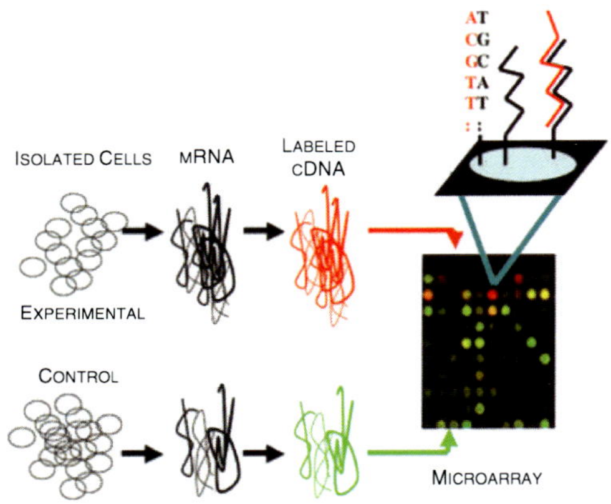

Plate 2.6 *Gene expression microarrays.* Microarrays can be used to measure the difference in relative expression between cells exposed to an experimental condition and a control condition. The isolated mRNA is labeled and hybridized to an array where each spot has reverse complement strands for a specific gene. If a gene is over-expressed in the experimental condition, the ratio of the experimental condition dye to the control condition dye at that spot will be much greater than one.

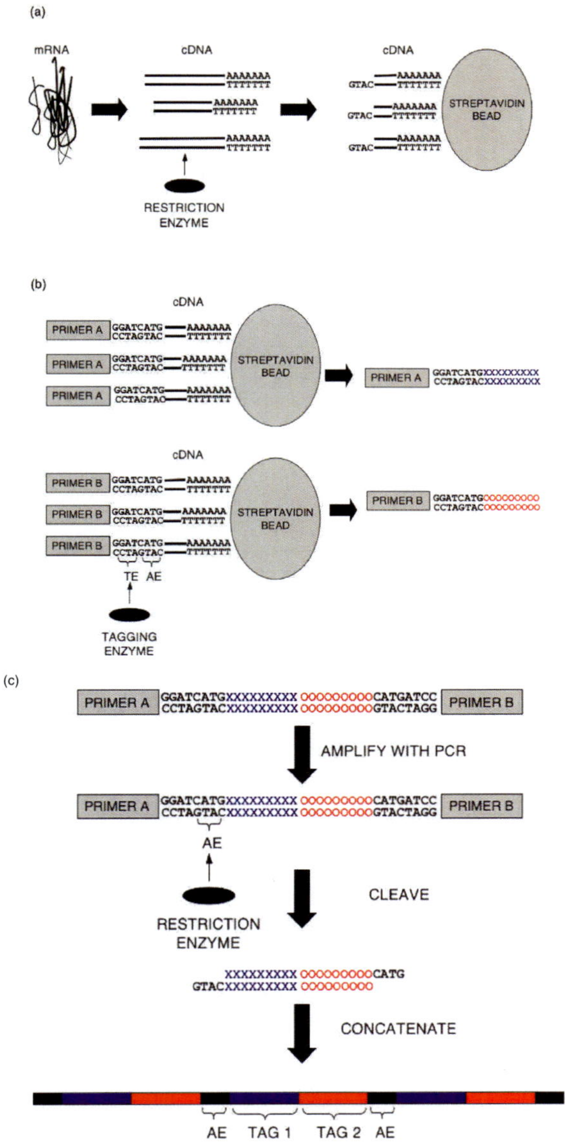

Plate 2.7 *Serial analysis of gene expression.* (a) To initiate gene expression measurement by SAGE, first mRNA is isolated. Reverse transcriptase is used to synthesize cDNA. A restriction enzyme, referred to as the "anchoring enzyme" is used. In this example, the anchoring enzyme recognizes and cleaves at the sequence "CATG" (AE). Beads are used to bind the poly-A tail of the cDNA fragments. (b) The beads are divided into two sets. Short sequences containing a PCR primer site and a "tagging enzyme" site (TE) are bound to the cDNA fragment ends. In this example the tagging enzyme recognizes "GGAT". Primer A is used in the first set, and Primer B is used in the second set. The tagging enzyme cleaves a specific distance downstream from its binding site. The remaining fragment contains a primer, the TE site, the AE site, and a unique sequence resulting from the original cDNA. (c) Sequence fragments from both sets are concatenated into ditags. The primers are then used in a PCR reaction to amplify the ditags. The anchoring enzyme is reused to cleave off the primer sequences and TE. The remaining sequence of the ditags can be concatenated into a long series of tags separated by AE sites. These longer sequences can be sequenced, and tags can then be easily identified.

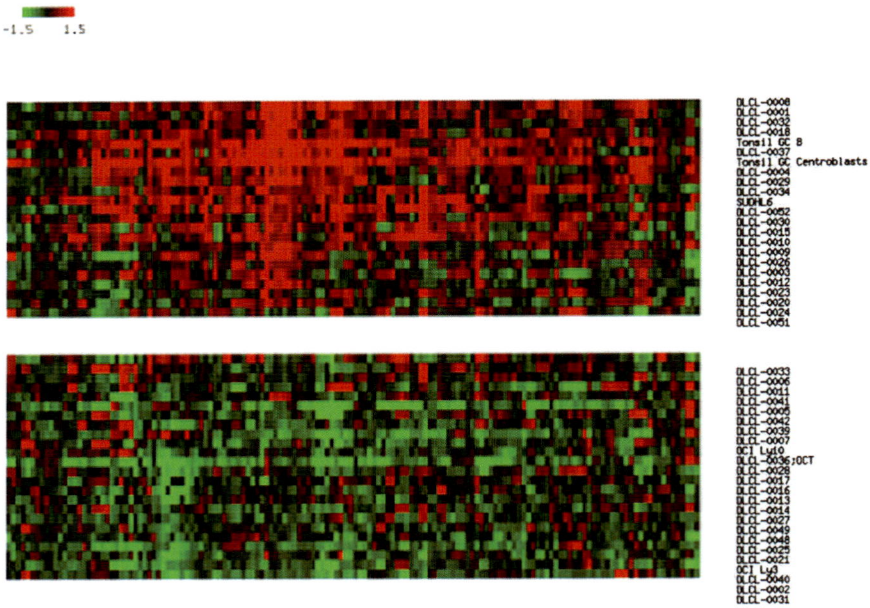

Plate 2.8 *K-means clustering of lymphoma data.* Lymphoma profiles were clustered using the expression of 148 germinal-center specific genes with a Euclidean distance metric. The resulting clusters are depicted in the graphic as two blocks. The upper block represents the germinal-cell subtype, and the lower block represents the activated subtype. Each column represents a specific gene, and each row a specific cancer profile.

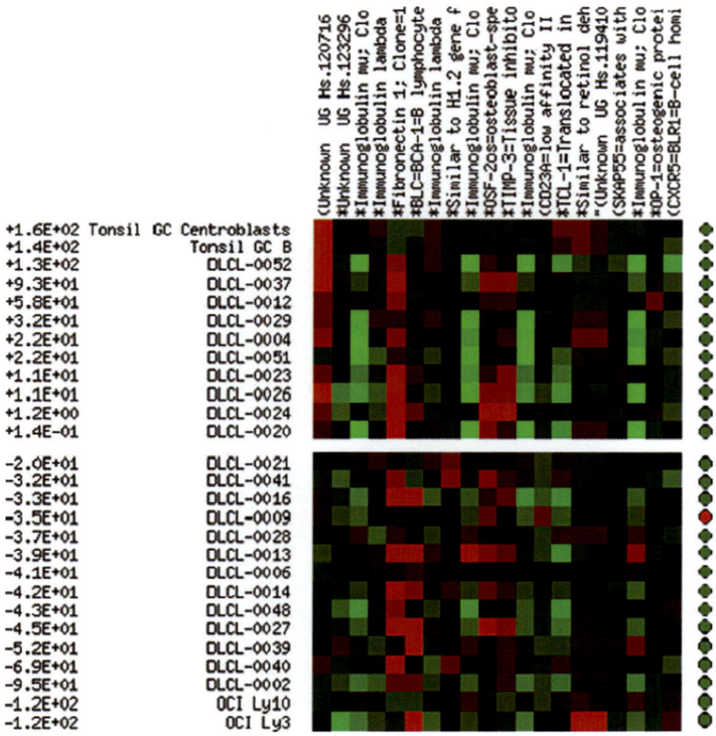

Plate 2.9 *Classification of lymphoma data into two classes with LDA.* Ten random profiles from each cluster defined in Plate 2.8 were defined as "known" examples. The remaining examples were compared with these to predict their class. A linear discriminant analysis classifier assigns log likelihood scores to each of the cases based on the expression of 4026 genes. Positive scores indicate membership in the germinal-center subtype, while negative scores indicate membership in the activated subtype. The larger the magnitude of the score, the more certain the prediction. The profiles are listed as rows in the graphic along with their confidence scores. The columns represent the genes that were given the greatest weight by the classifier. LDA assigns large weights to features most useful for distinguishing between labeled sets. Most cases were classified correctly (green dots); only one was classified incorrectly (red dot).

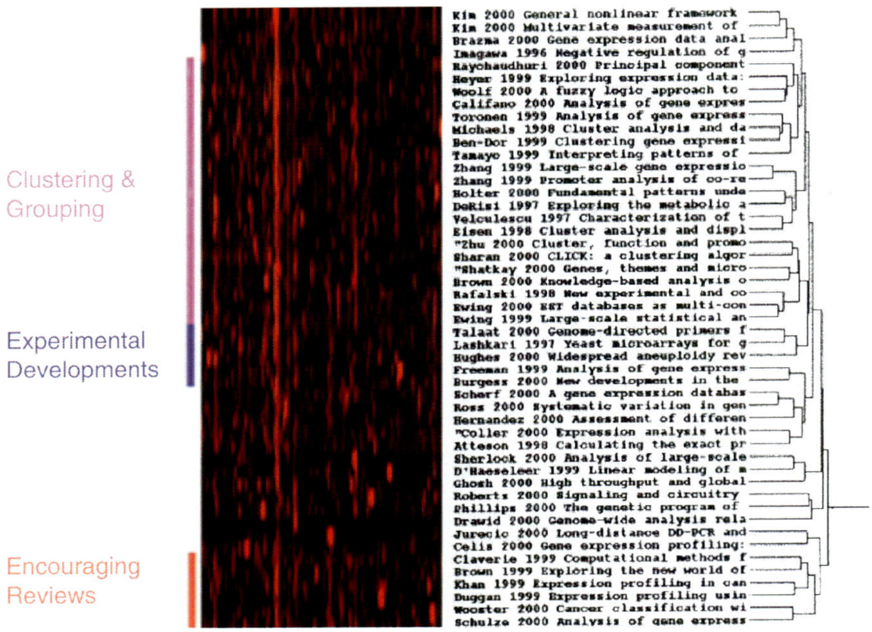

Plate 3.1 *Hierarchical clustering of gene expression analysis articles*. Here documents about gene expression analysis were converted into word vectors and clustered with a standard gene expression clustering package. Different classes of documents tend to cluster together as indicated by the bars and labels on the left.

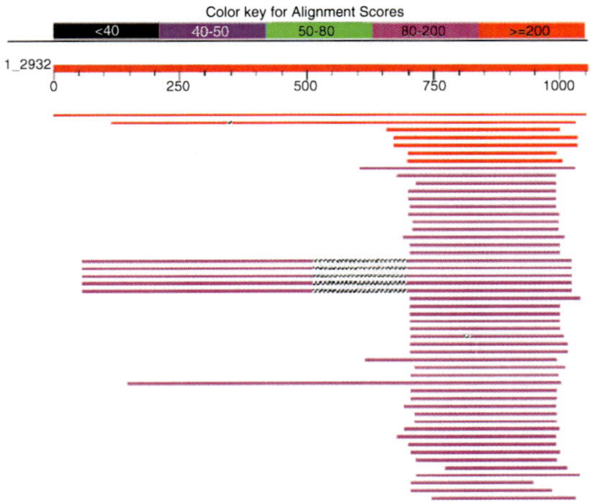

Plate 4.1 *Breathless* protein BLAST hits. BLAST hits tend to line up in a specific region of the *breathless* protein.

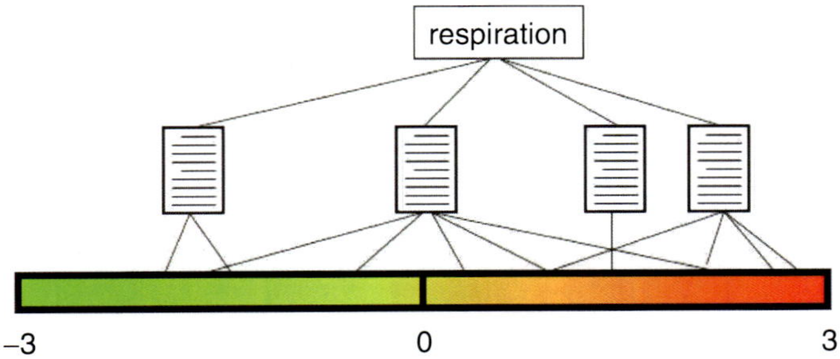

Plate 5.1 *Schematic for calculating "expression values" for a keyword.* Genes have a measured log expression ratio; the bar represents gene expression values. Average expression values for each article can be calculated by averaging the expression of the genes they refer to. The average expression for each word is the average of the expression of the articles they are contained in.

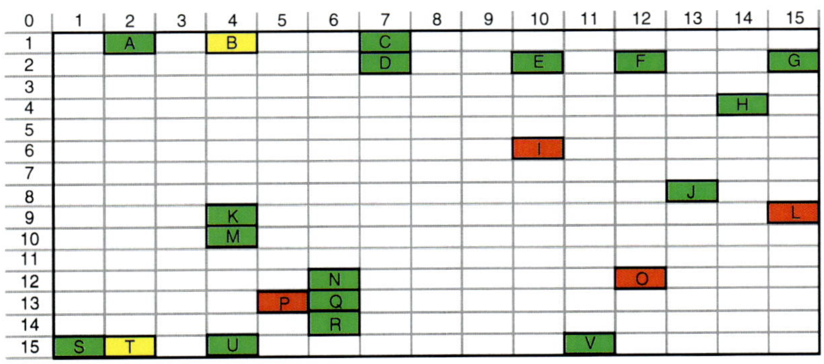

Plate 7.1 *Using NDPG to screen self-organizing map clusters.* Here we depict the self-organizing map clusters derived form the yeast data set. The clusters with more than five genes and NDPG scores greater than 3 are labeled with a letter. Clusters with NDPG scores greater than 3 are colored green, greater than 6 are colored yellow, and greater than 9 are colored red.

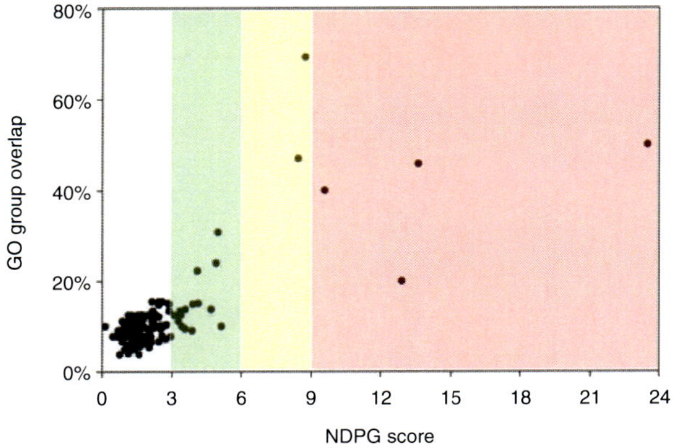

Plate 7.2 *Correlation between Gene Ontology Group overlap and NDPG score.* Here we show that there is a correlation between the NDPG score of a functional group and the percentage overlap the group has with a predefined Gene Ontology group. Both of these are indirect measures of a cluster's functional coherence. The green region indicates clusters with NDPG scores greater than 3, the yellow region indicates scores greater than 6, and red indicates scores greater than 9. Notice that an NDPG score less than 3 almost always implies little overlap with any Gene Ontology functional group.

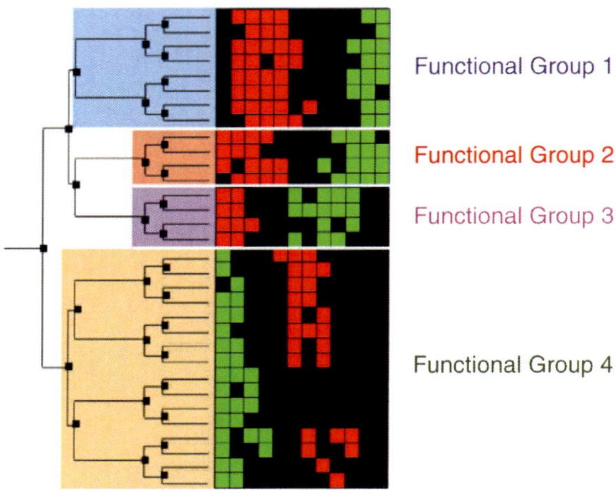

Plate 7.3 *Schematic of hierarchically clustered expression data with subsequent cluster boundary definition.* On the right is gene expression data represented as a colored grid. Each row in the grid represents the expression of a single gene across multiple conditions; each column represents the expression of the genes under those conditions. Red squares indicate gene induction while green squares indicate repression. On the left is a tree generated by a hierarchical clustering algorithm. The tree consists of nodes (dark boxes) that organize the genes according to expression similarity. All of the genes that descend from one node are the genes in the candidate cluster defined by that node. In this schematic we illustrate a pruning of the tree into four disjoint biologically relevant gene clusters. Pruning the tree defines concrete clusters and their boundaries.

After clustering the data, one must identify the biologically significant candidate clusters. Typically, careful expert examination of the genes in the clusters is required to identify the critical clusters in which the genes share function and to draw cluster boundaries that respect biological function. We assert that scientific literature can be mined automatically instead to identify biologically consistent clusters.

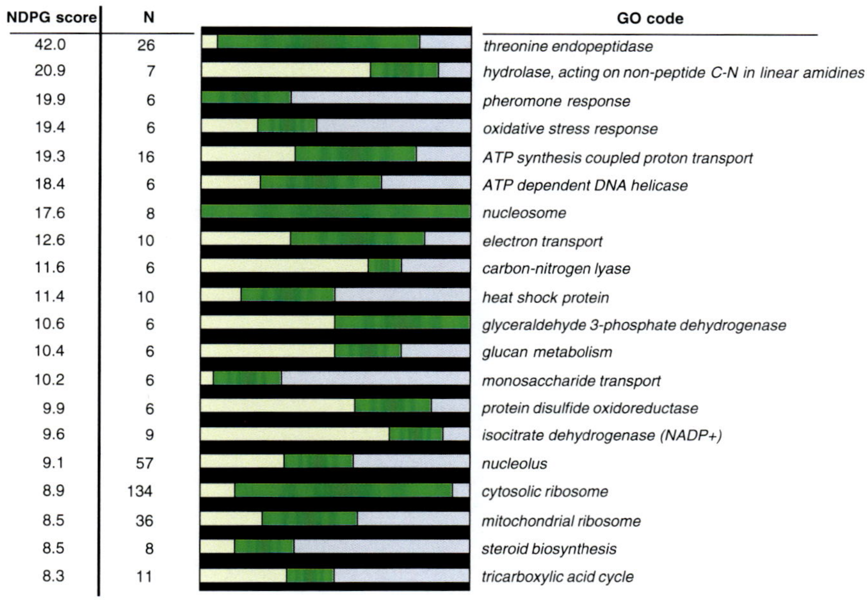

Plate 7.4 *Top 20 yeast gene clusters in order of NDPG scores.* These are the highest scoring clusters found after pruning. On the left of the graphic we list the *NDPG* score of each cluster and the number of genes within the cluster. On the right we list the GO code that best corresponds to the cluster. The length of the green bar in the graphic is proportional to the number of genes in the cluster that are also assigned the GO function listed on the right. The length of the yellow bar is proportional to the number of genes in the cluster not assigned the corresponding function by GO. The length of the blue bar is proportional to the number of additional genes assigned the GO function that are not in the cluster. The longer the green bar, the better the cluster represents that specific function.

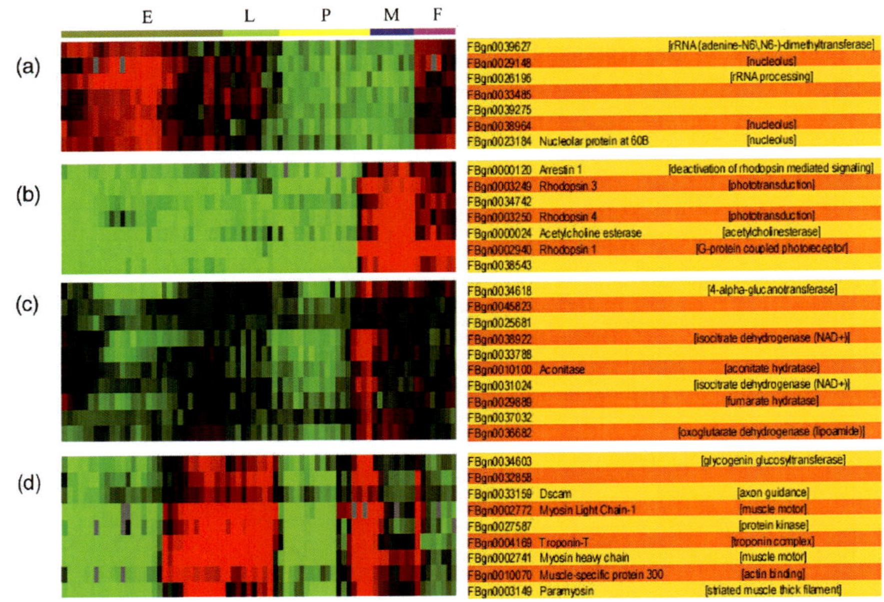

Plate 7.5 *Four gene expression clusters from the fly development data set whose boundaries were defined with the scientific literature.* The gene expression conditions are annotated at the top with E (embryo), L (larvae), P (pupae), M (adult male), and F (adult female). On the right genes are listed by FlyBase ID and name if available. On the far right we have listed the appropriate GO code annotation for that gene if available. (a) *Nucleolar maternal genes.* This cluster had not been identified in the original publication. (b) *Photoreceptor genes.* We found two separate photoreceptor clusters as did the authors of the original publication. (c) *Citric acid cycle genes.* Most of these genes have not been studied. Using sequence homology to obtain additional references made it feasible to identify this cluster of genes. A related but broader cluster was identified in the original publication. (d) *Muscle-specific genes.* A similar but broader cluster was identified in the original publication containing more unknown genes.